Dimensions Math®
Teacher's Guide 4A

Authors and Reviewers
Cassandra Turner
Allison Coates
Jenny Kempe
Bill Jackson
Tricia Salerno

Singapore Math Inc.

Published by Singapore Math Inc.

19535 SW 129th Avenue
Tualatin, OR 97062
www.singaporemath.com

Dimensions Math® Teacher's Guide 4A
ISBN 978-1-947226-38-8

First published 2019
Reprinted 2020 (twice), 2021

Copyright © 2019 by Singapore Math Inc.
All rights reserved. This book or any portion thereof may not be reproduced or used in any manner whatsoever without the express written permission of the publisher.

Printed in China

Acknowledgments

Editing by the Singapore Math Inc. team.
Design and illustration by Cameron Wray with Carli Bartlett.

Contents

Chapter		Lesson	Page
Chapter 1 **Numbers to One Million**		Teaching Notes	1
		Chapter Opener	5
	1	Numbers to 100,000	6
	2	Numbers to 1,000,000	10
	3	Number Patterns	14
	4	Comparing and Ordering Numbers	18
	5	Rounding 5-Digit Numbers	21
	6	Rounding 6-Digit Numbers	24
	7	Calculations and Place Value	26
	8	Practice	28
		Workbook Pages	30
Chapter 2 **Addition and Subtraction**		Teaching Notes	39
		Chapter Opener	45
	1	Addition	46
	2	Subtraction	49
	3	Other Ways to Add and Subtract — Part 1	52
	4	Other Ways to Add and Subtract — Part 2	54
	5	Word Problems	57
	6	Practice	59
		Workbook Pages	61
Chapter 3 **Multiples and Factors**		Teaching Notes	69
		Chapter Opener	73
	1	Multiples	74
	2	Common Multiples	77
	3	Factors	81
	4	Prime Numbers and Composite Numbers	84
	5	Common Factors	87
	6	Practice	90
		Workbook Pages	92

© 2019 Singapore Math Inc. Teacher's Guide 4A

Chapter		Lesson	Page
Chapter 4 **Multiplication**		Teaching Notes	99
		Chapter Opener	105
	1	Mental Math for Multiplication	106
	2	Multiplying by a 1-Digit Number — Part 1	110
	3	Multiplying by a 1-Digit Number — Part 2	113
	4	Practice A	116
	5	Multiplying by a Multiple of 10	119
	6	Multiplying by a 2-Digit Number — Part 1	123
	7	Multiplying by a 2-Digit Number — Part 2	125
	8	Practice B	127
		Workbook Pages	130
Chapter 5 **Division**		Teaching Notes	139
		Chapter Opener	145
	1	Mental Math for Division	146
	2	Estimation and Division	148
	3	Dividing 4-Digit Numbers	152
	4	Practice A	156
	5	Word Problems	158
	6	Challenging Word Problems	162
	7	Practice B	166
		Review 1	168
		Workbook Pages	170

Chapter		Lesson	Page
Chapter 6 **Fractions**		Teaching Notes	179
		Chapter Opener	183
	1	Equivalent Fractions	184
	2	Comparing and Ordering Fractions	187
	3	Improper Fractions and Mixed Numbers	190
	4	Practice A	192
	5	Expressing an improper Fraction as a Mixed Number	194
	6	Expressing a Mixed Number as an Improper Fraction	196
	7	Fractions and Division	198
	8	Practice B	202
		Workbook Pages	203
Chapter 7 **Adding and Subtracting Fractions**		Teaching Notes	211
		Chapter Opener	215
	1	Adding and Subtracting Fractions — Part 1	216
	2	Adding and Subtracting Fractions — Part 2	218
	3	Adding a Mixed Number and a Fraction	220
	4	Adding Mixed Numbers	223
	5	Subtracting a Fraction from a Mixed Number	225
	6	Subtracting Mixed Numbers	227
	7	Practice	230
		Workbook Pages	232

Chapter		Lesson	Page
Chapter 8 **Multiplying a Fraction and a Whole Number**		Teaching Notes	239
		Chapter Opener	243
	1	Multiplying a Unit Fraction by a Whole Number	244
	2	Multiplying a Fraction by a Whole Number — Part 1	246
	3	Multiplying a Fraction by a Whole Number — Part 2	248
	4	Fraction of a Set	250
	5	Multiplying a Whole Number by a Fraction — Part 1	253
	6	Multiplying a Whole Number by a Fraction — Part 2	257
	7	Word Problems — Part 1	260
	8	Word Problems — Part 2	263
	9	Practice	266
		Workbook Pages	267
Chapter 9 **Line Graphs and Line Plots**		Teaching Notes	277
		Chapter Opener	281
	1	Line Graphs	282
	2	Drawing Line Graphs	286
	3	Line Plots	288
	4	Practice	291
		Review 2	293
		Workbook Pages	296
Resources		Blackline Masters for 4A	301

Dimensions Math® Curriculum

The **Dimensions Math®** series is a Pre-Kindergarten to Grade 5 series based on the pedagogy and methodology of math education in Singapore. The main goal of the **Dimensions Math®** series is to help students develop competence and confidence in mathematics.

The series follows the principles outlined in the Singapore Mathematics Framework below.

Pedagogical Approach and Methodology

- Through Concrete-Pictorial-Abstract development, students view the same concepts over time with increasing levels of abstraction.
- Thoughtful sequencing creates a sense of continuity. The content of each grade level builds on that of preceding grade levels. Similarly, lessons build on previous lessons within each grade.
- Group discussion of solution methods encourages expansive thinking.
- Interesting problems and activities provide varied opportunities to explore and apply skills.
- Hands-on tasks and sharing establish a culture of collaboration.
- Extra practice and extension activities encourage students to persevere through challenging problems.
- Variation in pictorial representation (number bonds, bar models, etc.) and concrete representation (straws, linking cubes, base ten blocks, discs, etc.) broaden student understanding.

Each topic is introduced, then thoughtfully developed through the use of a variety of learning experiences, problem solving, student discourse, and opportunities for mastery of skills. This combination of hands-on practice, in-depth exploration of topics, and mathematical variability in teaching methodology allows students to truly master mathematical concepts.

Singapore Mathematics Framework

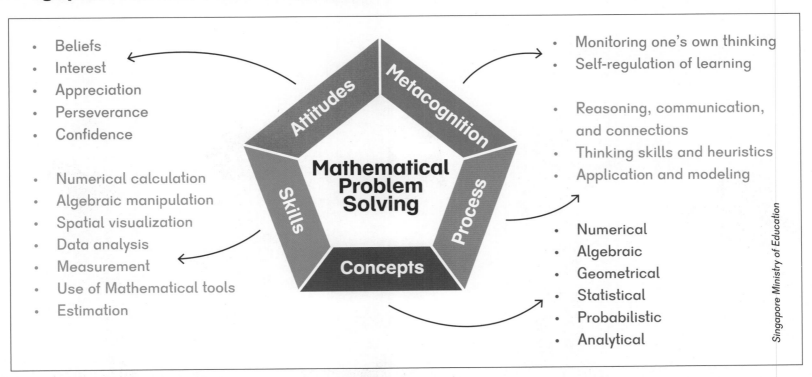

Dimensions Math® Program Materials

Textbooks

Textbooks are designed to help students build a solid foundation in mathematical thinking and efficient problem solving. Careful sequencing of topics, well-chosen problems, and simple graphics foster deep conceptual understanding and confidence. Mental math, problem solving, and correct computation are given balanced attention in all grades. As skills are mastered, students move to increasingly sophisticated concepts within and across grade levels.

Students work through the textbook lessons with the help of five friends: Emma, Alex, Sofia, Dion, and Mei. The characters appear throughout the series and help students develop metacognitive reasoning through questions, hints, and ideas.

A pencil icon ✏️ at the end of the textbook lessons links to exercises in the workbooks.

Workbooks

Workbooks provide additional problems that range from basic to challenging. These allow students to independently review and practice the skills they have learned.

Teacher's Guides

Teacher's Guides include lesson plans, mathematical background, games, helpful suggestions, and comprehensive resources for daily lessons.

Tests

Tests contain differentiated assessments to systematically evaluate student progress.

Emma   Alex   Sofia   Dion   Mei

Online Resources

The following can be downloaded from dimensionsmath.com.

- **Blackline Masters** used for various hands-on tasks.

- **Material Lists** for each chapter and lesson, so teachers and classroom helpers can prepare ahead of time.

- **Activities** that can be done with students who need more practice or a greater challenge, organized by concept, chapter, and lesson.

- **Standards Alignments** for various states.

Using the Teacher's Guide

This guide is designed to assist in planning daily lessons. It should be considered a helping hand between the curriculum and the classroom. It provides introductory notes on mathematical content, key points, and suggestions for activities. It also includes ideas for differentiation within each lesson, and answers and solutions to textbook and workbook problems.

Each chapter of the guide begins with the following.

- **Overview**

  Includes objectives and suggested number of class periods for each chapter.

- **Notes**

  Highlights key learning points, provides background on math concepts, explains the purpose of certain activities, and helps teachers understand the flow of topics throughout the year.

- **Materials**

  Lists materials, manipulatives, and Blackline Masters used in the Think and Learn sections of the guide. It also includes suggested storybooks. Many common classroom manipulatives are used throughout the curriculum. When a lesson refers to a whiteboard and markers, any writing materials can be used. Blackline Masters can be found at dimensionsmath.com.

The guide goes through the Chapter Openers, Daily Lessons, and Practices of each chapter, and cumulative reviews in the following general format.

- **Chapter Opener**

  Provides talking points for discussion to prepare students for the math concepts to be introduced.

- **Think**

  Offers structure for teachers to guide student inquiry. Provides various methods and activities to solve initial textbook problems or tasks.

- **Learn**

  Guides teachers to analyze student methods from Think to arrive at the main concepts of the lesson through discussion and study of the pictorial representations in the textbook.

- **Do**

  Expands on specific problems with strategies, additional practice, and remediation.

● Activities

Allows students to practice concepts through individual, small group, and whole group hands-on tasks and games, including suggestions for outdoor play (most of which can be modified for a gymnasium or classroom).

Level of difficulty in the games and activities are denoted by the following symbols.

- ● Foundational activities
- ▲ On-level activities
- ★ Challenge or extension activities

● Brain Works

Provides opportunities for students to extend their mathematical thinking.

Discussion is a critical component of each lesson. Teachers are encouraged to let students discuss their reasoning. As each classroom is different, this guide does not anticipate all situations. The following questions can help students articulate their thinking and increase their mastery:

- Why? How do you know?
- Can you explain that?
- Can you draw a picture of that?
- Is your answer reasonable? How do you know?
- How is this task like the one we did before? How is it different?
- What is alike and what is different about…?
- Can you solve that a different way?
- Yes! You're right! How do you know it's true?
- What did you learn before that can help you solve this problem?
- Can you summarize what your classmate shared?
- What conclusion can you draw from the data?

Each lesson is designed to take one day. If your calendar allows, you may choose to spend more than one day on certain lessons. Throughout the guide, there are notes to extend on learning activities to make them more challenging. Lesson structures and activities do not have to conform exactly to what is shown in the guide. Teachers are encouraged to exercise their discretion in using this material in a way that best suits their classes.

Textbooks are designed to last multiple years. Textbook problems with a ▇ (or a blank line for terms) are meant to invite active participation.

Dimensions Math® Scope & Sequence

PKA

Chapter 1
Match, Sort, and Classify

Red and Blue
Yellow and Green
Color Review
Soft and Hard
Rough, Bumpy, and Smooth
Sticky and Grainy
Size — Part 1
Size — Part 2
Sort Into Two Groups
Practice

Chapter 2
Compare Objects

Big and Small
Long and Short
Tall and Short
Heavy and Light
Practice

Chapter 3
Patterns

Movement Patterns
Sound Patterns
Create Patterns
Practice

Chapter 4
Numbers to 5 — Part 1

Count 1 to 5 — Part 1
Count 1 to 5 — Part 2
Count Back

Count On and Back
Count 1 Object
Count 2 Objects
Count Up to 3 Objects
Count Up to 4 Objects
Count Up to 5 Objects
How Many? — Part 1
How Many? — Part 2
How Many Now? — Part 1
How Many Now? — Part 2
Practice

Chapter 5
Numbers to 5 — Part 2

1, 2, 3
1, 2, 3, 4, 5 — Part 1
1, 2, 3, 4, 5 — Part 2
How Many? — Part 1
How Many? — Part 2
How Many Do You See?
How Many Do You See Now?
Practice

Chapter 6
Numbers to 10 — Part 1

0
Count to 10 — Part 1
Count to 10 — Part 2
Count Back
Order Numbers
Count Up to 6 Objects
Count Up to 7 Objects
Count Up to 8 Objects
Count Up to 9 Objects
Count Up to 10 Objects — Part 1

Count Up to 10 Objects — Part 2
How Many?
Practice

Chapter 7
Numbers to 10 — Part 2

6
7
8
9
10
0 to 10
Count and Match — Part 1
Count and Match — Part 2
Practice

PKB

Chapter 8
Ordinal Numbers

First
Second and Third
Fourth and Fifth
Practice

Chapter 9
Shapes and Solids

Cubes, Cylinders, and Spheres
Cubes
Positions
Build with Solids
Rectangles and Circles
Squares
Triangles

Squares, Circles, Rectangles, and Triangles — Part 1
Squares, Circles, Rectangles, and Triangles — Part 2
Practice

Chapter 10
Compare Sets

Match Objects
Which Set Has More?
Which Set Has Fewer?
More or Fewer?
Practice

Chapter 11
Compose and Decompose

Altogether — Part 1
Altogether — Part 2
Show Me
What's the Other Part? — Part 1
What's the Other Part? — Part 2
Practice

Chapter 12
Explore Addition and Subtraction

Add to 5 — Part 1
Add to 5 — Part 2
Two Parts Make a Whole
How Many in All?
Subtract Within 5 — Part 1
Subtract Within 5 — Part 2
How Many Are Left?

Practice

Chapter 13
Cumulative Review

Review 1 Match and Color
Review 2 Big and Small
Review 3 Heavy and Light
Review 4 Count to 5
Review 5 Count 5 Objects
Review 6 0
Review 7 Count Beads
Review 8 Patterns
Review 9 Length
Review 10 How Many?
Review 11 Ordinal Numbers
Review 12 Solids and Shapes
Review 13 Which Set Has More?
Review 14 Which Set Has Fewer?
Review 15 Put Together
Review 16 Subtraction
Looking Ahead 1 Sequencing — Part 1
Looking Ahead 2 Sequencing — Part 2
Looking Ahead 3 Categorizing
Looking Ahead 4 Addition
Looking Ahead 5 Subtraction
Looking Ahead 6 Getting Ready to Write Numerals
Looking Ahead 7 Reading and Math

KA

Chapter 1
Match, Sort, and Classify

Left and Right
Same and Similar
Look for One That Is Different
How Does it Feel?
Match the Things That Go Together
Sort
Practice

Chapter 2
Numbers to 5

Count to 5
Count Things Up to 5
Recognize the Numbers 1 to 3
Recognize the Numbers 4 and 5
Count and Match
Write the Numbers 1 and 2
Write the Number 3
Write the Number 4
Trace and Write 1 to 5
Zero
Picture Graphs
Practice

Chapter 3
Numbers to 10

Count 1 to 10
Count Up to 7 Things
Count Up to 9 Things
Count Up to 10 Things — Part 1

Dimensions Math® Scope & Sequence

Count Up to 10 Things —
   Part 2
Recognize the Numbers
   6 to 10
Write the Numbers 6 and 7
Write the Numbers 8, 9,
   and 10
Write the Numbers 6 to 10
Count and Write the
   Numbers 1 to 10
Ordinal Positions
One More Than
Practice

Chapter 4
Shapes and Solids

Curved or Flat
Solid Shapes
Closed Shapes
Rectangles
Squares
Circles and Triangles
Where is It?
Hexagons
Sizes and Shapes
Combine Shapes
Graphs
Practice

Chapter 5
Compare Height, Length, Weight, and Capacity

Comparing Height
Comparing Length
Height and Length — Part 1
Height and Length — Part 2
Weight — Part 1
Weight — Part 2
Weight — Part 3
Capacity — Part 1
Capacity — Part 2
Practice

Chapter 6
Comparing Numbers Within 10

Same and More
More and Fewer
More and Less
Practice — Part 1
Practice — Part 2

KB

Chapter 7
Numbers to 20

Ten and Some More
Count Ten and Some More
Two Ways to Count
Numbers 16 to 20
Number Words 0 to 10
Number Words 11 to 15
Number Words 16 to 20
Number Order
1 More Than or Less Than
Practice — Part 1
Practice — Part 2

Chapter 8
Number Bonds

Putting Numbers Together
   — Part 1
Putting Numbers Together
   — Part 2
Parts Making a Whole
Look for a Part
Number Bonds for 2, 3, and 4
Number Bonds for 5
Number Bonds for 6
Number Bonds for 7
Number Bonds for 8
Number Bonds for 9
Number Bonds for 10
Practice — Part 1
Practice — Part 2
Practice — Part 3

Chapter 9
Addition

Introduction to Addition —
   Part 1
Introduction to Addition —
   Part 2
Introduction to Addition —
   Part 3
Addition
Count On — Part 1
Count On — Part 2
Add Up to 3 and 4
Add Up to 5 and 6
Add Up to 7 and 8
Add Up to 9 and 10
Addition Practice
Practice

Chapter 10
Subtraction

Take Away to Subtract —
   Part 1

Take Away to Subtract — Part 2
Take Away to Subtract — Part 3
Take Apart to Subtract — Part 1
Take Apart to Subtract — Part 2
Count Back
Subtract Within 5
Subtract Within 10 — Part 1
Subtract Within 10 — Part 2
Practice

Chapter 11
Addition and Subtraction

Add and Subtract
Practice Addition and Subtraction
Part-Whole Addition and Subtraction
Add to or Take Away
Put Together or Take Apart
Practice

Chapter 12
Numbers to 100

Count by Tens — Part 1
Count by Tens — Part 2
Numbers to 30
Numbers to 40
Numbers to 50
Numbers to 80
Numbers to 100 — Part 1
Numbers to 100 — Part 2
Count by Fives — Part 1
Count by Fives — Part 2

Practice

Chapter 13
Time

Day and Night
Learning About the Clock
Telling Time to the Hour — Part 1
Telling Time to the Hour — Part 2
Practice

Chapter 14
Money

Coins
Pennies
Nickels
Dimes
Quarters
Practice

1A

Chapter 1
Numbers to 10

Numbers to 10
The Number 0
Order Numbers
Compare Numbers
Practice

Chapter 2
Number Bonds

Make 6
Make 7
Make 8
Make 9
Make 10 — Part 1
Make 10 — Part 2
Practice

Chapter 3
Addition

Addition as Putting Together
Addition as Adding More
Addition with 0
Addition with Number Bonds
Addition by Counting On
Make Addition Stories
Addition Facts
Practice

Chapter 4
Subtraction

Subtraction as Taking Away
Subtraction as Taking Apart
Subtraction by Counting Back
Subtraction with 0
Make Subtraction Stories
Subtraction with Number Bonds
Addition and Subtraction
Make Addition and Subtraction Story Problems
Subtraction Facts
Practice
Review 1

Chapter 5
Numbers to 20

Numbers to 20
Add or Subtract Tens or Ones
Order Numbers to 20

Dimensions Math® Scope & Sequence

Compare Numbers to 20
Addition
Subtraction
Practice

Chapter 6
Addition to 20

Add by Making 10 — Part 1
Add by Making 10 — Part 2
Add by Making 10 — Part 3
Addition Facts to 20
Practice

Chapter 7
Subtraction Within 20

Subtract from 10 — Part 1
Subtract from 10 — Part 2
Subtract the Ones First
Word Problems
Subtraction Facts Within 20
Practice

Chapter 8
Shapes

Solid and Flat Shapes
Grouping Shapes
Making Shapes
Practice

Chapter 9
Ordinal Numbers

Naming Positions
Word Problems
Practice
Review 2

1B

Chapter 10
Length

Comparing Lengths Directly
Comparing Lengths Indirectly
Comparing Lengths with Units
Practice

Chapter 11
Comparing

Subtraction as Comparison
Making Comparison Subtraction Stories
Picture Graphs
Practice

Chapter 12
Numbers to 40

Numbers to 40
Tens and Ones
Counting by Tens and Ones
Comparing
Practice

Chapter 13
Addition and Subtraction Within 40

Add Ones
Subtract Ones
Make the Next Ten
Use Addition Facts
Subtract from Tens
Use Subtraction Facts
Add Three Numbers
Practice

Chapter 14
Grouping and Sharing

Adding Equal Groups
Sharing
Grouping
Practice

Chapter 15
Fractions

Halves
Fourths
Practice
Review 3

Chapter 16
Numbers to 100

Numbers to 100
Tens and Ones
Count by Ones or Tens
Compare Numbers to 100
Practice

Chapter 17
Addition and Subtraction Within 100

Add Ones — Part 1
Add Tens
Add Ones — Part 2
Add Tens and Ones — Part 1
Add Tens and Ones — Part 2
Subtract Ones — Part 1
Subtract from Tens
Subtract Ones — Part 2
Subtract Tens

Subtract Tens and Ones — Part 1
Subtract Tens and Ones — Part 2
Practice

Chapter 18
Time

Telling Time to the Hour
Telling Time to the Half Hour
Telling Time to the 5 Minutes
Practice

Chapter 19
Money

Coins
Counting Money
Bills
Shopping
Practice
Review 4

2A

Chapter 1
Numbers to 1,000

Tens and Ones
Counting by Tens or Ones
Comparing Tens and Ones
Hundreds, Tens, and Ones
Place Value
Comparing Hundreds, Tens, and Ones
Counting by Hundreds, Tens, or Ones
Practice

Chapter 2
Addition and Subtraction — Part 1

Strategies for Addition
Strategies for Subtraction
Parts and Whole
Comparison
Practice

Chapter 3
Addition and Subtraction — Part 2

Addition Without Regrouping
Subtraction Without Regrouping
Addition with Regrouping Ones
Addition with Regrouping Tens
Addition with Regrouping Tens and Ones
Practice A
Subtraction with Regrouping from Tens
Subtraction with Regrouping from Hundreds
Subtraction with Regrouping from Two Places
Subtraction with Regrouping across Zeros
Practice B
Practice C

Chapter 4
Length

Centimeters
Estimating Length in Centimeters
Meters
Estimating Length in Meters
Inches
Using Rulers
Feet
Practice

Chapter 5
Weight

Grams
Kilograms
Pounds
Practice
Review 1

Chapter 6
Multiplication and Division

Multiplication — Part 1
Multiplication — Part 2
Practice A
Division — Part 1
Division — Part 2
Multiplication and Division
Practice B

Chapter 7
Multiplication and Division of 2, 5, and 10

The Multiplication Table of 5
Multiplication Facts of 5
Practice A
The Multiplication Table of 2
Multiplication Facts of 2
Practice B
The Multiplication Table of 10
Dividing by 2

Dimensions Math® Scope & Sequence

Dividing by 5 and 10
Practice C
Word Problems
Review 2

2B

Chapter 8
Mental Calculation

Adding Ones Mentally
Adding Tens Mentally
Making 100
Adding 97, 98, or 99
Practice A
Subtracting Ones Mentally
Subtracting Tens Mentally
Subtracting 97, 98, or 99
Practice B
Practice C

Chapter 9
Multiplication and Division of 3 and 4

The Multiplication Table of 3
Multiplication Facts of 3
Dividing by 3
Practice A
The Multiplication Table of 4
Multiplication Facts of 4
Dividing by 4
Practice B
Practice C

Chapter 10
Money

Making $1
Dollars and Cents
Making Change
Comparing Money
Practice A
Adding Money
Subtracting Money
Practice B

Chapter 11
Fractions

Halves and Fourths
Writing Unit Fractions
Writing Fractions
Fractions that Make 1 Whole
Comparing and Ordering Fractions
Practice
Review 3

Chapter 12
Time

Telling Time
Time Intervals
A.M. and P.M.
Practice

Chapter 13
Capacity

Comparing Capacity
Units of Capacity
Practice

Chapter 14
Graphs

Picture Graphs
Bar Graphs
Practice

Chapter 15
Shapes

Straight and Curved Sides
Polygons
Semicircles and Quarter-circles
Patterns
Solid Shapes
Practice
Review 4
Review 5

3A

Chapter 1
Numbers to 10,000

Numbers to 10,000
Place Value — Part 1
Place Value — Part 2
Comparing Numbers
The Number Line
Practice A
Number Patterns
Rounding to the Nearest Thousand
Rounding to the Nearest Hundred
Rounding to the Nearest Ten
Practice B

Chapter 2
Addition and Subtraction — Part 1

Mental Addition — Part 1
Mental Addition — Part 2
Mental Subtraction — Part 1
Mental Subtraction — Part 2
Making 100 and 1,000
Strategies for Numbers Close to Hundreds
Practice A
Sum and Difference
Word Problems — Part 1
Word Problems — Part 2
2-Step Word Problems
Practice B

Chapter 3
Addition and Subtraction — Part 2

Addition with Regrouping
Subtraction with Regrouping — Part 1
Subtraction with Regrouping — Part 2
Estimating Sums and Differences — Part 1
Estimating Sums and Differences — Part 2
Word Problems
Practice

Chapter 4
Multiplication and Division

Looking Back at Multiplication
Strategies for Finding the Product
Looking Back at Division
Multiplying and Dividing with 0 and 1
Division with Remainders
Odd and Even Numbers
Word Problems — Part 1
Word Problems — Part 2
2-Step Word Problems
Practice
Review 1

Chapter 5
Multiplication

Multiplying Ones, Tens, and Hundreds
Multiplication Without Regrouping
Multiplication with Regrouping Tens
Multiplication with Regrouping Ones
Multiplication with Regrouping Ones and Tens
Practice A
Multiplying a 3-Digit Number with Regrouping Once
Multiplication with Regrouping More Than Once
Practice B

Chapter 6
Division

Dividing Tens and Hundreds
Dividing a 2-Digit Number by 2 — Part 1
Dividing a 2-Digit Number by 2 — Part 2
Dividing a 2-Digit Number by 3, 4, and 5
Practice A
Dividing a 3-Digit Number by 2
Dividing a 3-Digit Number by 3, 4, and 5
Dividing a 3-Digit Number, Quotient is 2 Digits
Practice B

Chapter 7
Graphs and Tables

Picture Graphs and Bar Graphs
Bar Graphs and Tables
Practice
Review 2

3B

Chapter 8
Multiplying and Dividing with 6, 7, 8, and 9

The Multiplication Table of 6
The Multiplication Table of 7
Multiplying by 6 and 7
Dividing by 6 and 7
Practice A
The Multiplication Table of 8

Dimensions Math® Scope & Sequence

The Multiplication Table of 9
Multiplying by 8 and 9
Dividing by 8 and 9
Practice B

Chapter 9
Fractions — Part 1

Fractions of a Whole
Fractions on a Number Line
Comparing Fractions with Like Denominators
Comparing Fractions with Like Numerators
Practice

Chapter 10
Fractions — Part 2

Equivalent Fractions
Finding Equivalent Fractions
Simplifying Fractions
Comparing Fractions — Part 1
Comparing Fractions — Part 2
Practice A
Adding and Subtracting Fractions — Part 1
Adding and Subtracting Fractions — Part 2
Practice B

Chapter 11
Measurement

Meters and Centimeters
Subtracting from Meters
Kilometers
Subtracting from Kilometers
Liters and Milliliters
Kilograms and Grams

Word Problems
Practice
Review 3

Chapter 12
Geometry

Circles
Angles
Right Angles
Triangles
Properties of Triangles
Properties of Quadrilaterals
Using a Compass
Practice

Chapter 13
Area and Perimeter

Area
Units of Area
Area of Rectangles
Area of Composite Figures
Practice A
Perimeter
Perimeter of Rectangles
Area and Perimeter
Practice B

Chapter 14
Time

Units of Time
Calculating Time — Part 1
Practice A
Calculating Time — Part 2
Calculating Time — Part 3
Calculating Time — Part 4
Practice B

Chapter 15
Money

Dollars and Cents
Making $10
Adding Money
Subtracting Money
Word Problems
Practice
Review 4
Review 5

4A

Chapter 1
Numbers to One Million

Numbers to 100,000
Numbers to 1,000,000
Number Patterns
Comparing and Ordering Numbers
Rounding 5-Digit Numbers
Rounding 6-Digit Numbers
Calculations and Place Value
Practice

Chapter 2
Addition and Subtraction

Addition
Subtraction
Other Ways to Add and Subtract — Part 1
Other Ways to Add and Subtract — Part 2
Word Problems

Practice

Chapter 3
Multiples and Factors

Multiples
Common Multiples
Factors
Prime Numbers and
    Composite Numbers
Common Factors
Practice

Chapter 4
Multiplication

Mental Math for Multiplication
Multiplying by a 1-Digit
    Number — Part 1
Multiplying by a 1-Digit
    Number — Part 2
Practice A
Multiplying by a Multiple of 10
Multiplying by a 2-Digit
    Number — Part 1
Multiplying by a 2-Digit
    Number — Part 2
Practice B

Chapter 5
Division

Mental Math for Division
Estimation and Division
Dividing 4-Digit Numbers
Practice A
Word Problems
Challenging Word Problems
Practice B
Review 1

Chapter 6
Fractions

Equivalent Fractions
Comparing and Ordering
    Fractions
Improper Fractions and Mixed
    Numbers
Practice A
Expressing an Improper
    Fraction as a Mixed
    Number
Expressing a Mixed Number
    as an Improper Fraction
Fractions and Division
Practice B

Chapter 7
Adding and Subtracting Fractions

Adding and Subtracting
    Fractions — Part 1
Adding and Subtracting
    Fractions — Part 2
Adding a Mixed Number and
    a Fraction
Adding Mixed Numbers
Subtracting a Fraction from
    a Mixed Number
Subtracting Mixed Numbers
Practice

Chapter 8
Multiplying a Fraction and a Whole Number

Multiplying a Unit Fraction
    by a Whole Number
Multiplying a Fraction by a
    Whole Number — Part 1
Multiplying a Fraction by a
    Whole Number — Part 2
Fraction of a Set
Multiplying a Whole Number
    by a Fraction — Part 1
Multiplying a Whole Number
    by a Fraction — Part 2
Word Problems — Part 1
Word Problems — Part 2
Practice

Chapter 9
Line Graphs and Line Plots

Line Graphs
Drawing Line Graphs
Line Plots
Practice
Review 2

4B

Chapter 10
Measurement

Metric Units of Measurement
Customary Units of Length
Customary Units of Weight
Customary Units of Capacity
Units of Time
Practice A
Fractions and Measurement
    — Part 1
Fractions and Measurement
    — Part 2
Practice B

Dimensions Math® Scope & Sequence

Chapter 11
Area and Perimeter

Area of Rectangles — Part 1
Area of Rectangles — Part 2
Area of Composite Figures
Perimeter — Part 1
Perimeter — Part 2
Practice

Chapter 12
Decimals

Tenths — Part 1
Tenths — Part 2
Hundredths — Part 1
Hundredths — Part 2
Expressing Decimals as Fractions in Simplest Form
Expressing Fractions as Decimals
Practice A
Comparing and Ordering Decimals
Rounding Decimals
Practice B

Chapter 13
Addition and Subtraction of Decimals

Adding and Subtracting Tenths
Adding Tenths with Regrouping
Subtracting Tenths with Regrouping
Practice A
Adding Hundredths
Subtracting from 1 and 0.1
Subtracting Hundredths
Money, Decimals, and Fractions

Practice B
Review 3

Chapter 14
Multiplication and Division of Decimals

Multiplying Tenths and Hundredths
Multiplying Decimals by a Whole Number — Part 1
Multiplying Decimals by a Whole Number — Part 2
Practice A
Dividing Tenths and Hundredths
Dividing Decimals by a Whole Number — Part 1
Dividing Decimals by a Whole Number — Part 2
Dividing Decimals by a Whole Number — Part 3
Practice B

Chapter 15
Angles

The Size of Angles
Measuring Angles
Drawing Angles
Adding and Subtracting Angles
Reflex Angles
Practice

Chapter 16
Lines and Shapes

Perpendicular Lines
Parallel Lines
Drawing Perpendicular and Parallel Lines
Quadrilaterals

Lines of Symmetry
Symmetrical Figures and Patterns
Practice

Chapter 17
Properties of Cuboids

Cuboids
Nets of Cuboids
Faces and Edges of Cuboids
Practice
Review 4
Review 5

5A

Chapter 1
Whole Numbers

Numbers to One Billion
Multiplying by 10, 100, and 1,000
Dividing by 10, 100, and 1,000
Multiplying by Tens, Hundreds, and Thousands
Dividing by Tens, Hundreds, and Thousands
Practice

Chapter 2
Writing and Evaluating Expressions

Expressions with Parentheses
Order of Operations — Part 1
Order of Operations — Part 2

Other Ways to Write and Evaluate Expressions
Word Problems — Part 1
Word Problems — Part 2
Practice

Chapter 3
Multiplication and Division

Multiplying by a 2-digit Number — Part 1
Multiplying by a 2-digit Number — Part 2
Practice A
Dividing by a Multiple of Ten
Divide a 2-digit Number by a 2-digit Number
Divide a 3-digit Number by a 2-digit Number — Part 1
Divide a 3-digit Number by a 2-digit Number — Part 2
Divide a 4-digit Number by a 2-digit Number
Practice B

Chapter 4
Addition and Subtraction of Fractions

Fractions and Division
Adding Unlike Fractions
Subtracting Unlike Fractions
Practice A
Adding Mixed Numbers — Part 1
Adding Mixed Numbers — Part 2
Subtracting Mixed Numbers — Part 1

Subtracting Mixed Numbers — Part 2
Practice B
Review 1

Chapter 5
Multiplication of Fractions

Multiplying a Fraction by a Whole Number
Multiplying a Whole Number by a Fraction
Word Problems — Part 1
Practice A
Multiplying a Fraction by a Unit Fraction
Multiplying a Fraction by a Fraction — Part 1
Multiplying a Fraction by a Fraction — Part 2
Multiplying Mixed Numbers
Word Problems — Part 2
Fractions and Reciprocals
Practice B

Chapter 6
Division of Fractions

Dividing a Unit Fraction by a Whole Number
Dividing a Fraction by a Whole Number
Practice A
Dividing a Whole Number by a Unit Fraction
Dividing a Whole Number by a Fraction
Word Problems
Practice B

Chapter 7
Measurement

Fractions and Measurement Conversions
Fractions and Area
Practice A
Area of a Triangle — Part 1
Area of a Triangle — Part 2
Area of Complex Figures
Practice B

Chapter 8
Volume of Solid Figures

Cubic Units
Volume of Cuboids
Finding the Length of an Edge
Practice A
Volume of Complex Shapes
Volume and Capacity — Part 1
Volume and Capacity — Part 2
Practice B
Review 2

5B

Chapter 9
Decimals

Thousandths
Place Value to Thousandths
Comparing Decimals
Rounding Decimals
Practice A
Multiply Decimals by 10, 100, and 1,000
Divide Decimals by 10, 100, and 1,000

Dimensions Math® Scope & Sequence

Conversion of Measures
Mental Calculation
Practice B

Chapter 10
The Four Operations of Decimals

Adding Decimals to Thousandths
Subtracting Decimals
Multiplying by 0.1 or 0.01
Multiplying by a Decimal
Practice A
Dividing by a Whole Number — Part 1
Dividing by a Whole Number — Part 2
Dividing a Whole Number by 0.1 and 0.01
Dividing a Whole Number by a Decimal
Practice B

Chapter 11
Geometry

Measuring Angles
Angles and Lines
Classifying Triangles
The Sum of the Angles in a Triangle
The Exterior Angle of a Triangle
Classifying Quadrilaterals
Angles of Quadrilaterals — Part 1
Angles of Quadrilaterals — Part 2
Drawing Triangles and Quadrilaterals
Practice

Chapter 12
Data Analysis and Graphs

Average — Part 1
Average — Part 2
Line Plots
Coordinate Graphs
Straight Line Graphs
Practice
Review 3

Chapter 13
Ratio

Finding the Ratio
Equivalent Ratios
Finding a Quantity
Comparing Three Quantities
Word Problems
Practice

Chapter 14
Rate

Finding the Rate
Rate Problems — Part 1
Rate Problems — Part 2
Word Problems
Practice

Chapter 15
Percentage

Meaning of Percentage
Expressing Percentages as Fractions
Percentages and Decimals
Expressing Fractions as Percentages
Practice A
Percentage of a Quantity
Word Problems
Practice B
Review 4
Review 5

Chapter 1 Numbers to One Million

Overview

Suggested number of class periods: 8–9

Lesson		Page	Resources	Objectives
	Chapter Opener	p. 5	TB: p. 1	Investigate large numbers.
1	Numbers to 100,000	p. 6	TB: p. 2 WB: p. 1	Understand how numbers to 100,000 are composed based on place value. Write up to five-digit numbers using expanded form.
2	Numbers to 1,000,000	p. 10	TB: p. 7 WB: p. 4	Understand how numbers to 1,000,000 are composed based on place value. Write up to six-digit numbers using expanded form.
3	Number Patterns	p. 14	TB: p. 12 WB: p. 7	Find a number that is 100, 1,000, or 10,000 more or less than a five-digit number. Count on and back from a given number by hundreds, thousands, or ten thousands.
4	Comparing and Ordering Numbers	p. 18	TB: p. 18 WB: p. 10	Compare and order numbers within 1 million.
5	Rounding 5-digit Numbers	p. 21	TB: p. 22 WB: p. 13	Round a five-digit number to the nearest ten thousand, thousand, or hundred.
6	Rounding 6-digit Numbers	p. 24	TB: p. 26 WB: p. 16	Round a six-digit number to the nearest hundred thousand, ten thousand, or thousand.
7	Calculations and Place Value	p. 26	TB: p. 30 WB: p. 19	Use place value concepts in mental computations.
8	Practice	p. 28	TB: p. 33 WB: p. 22	Practice concepts from the chapter.
	Workbook Solutions	p. 30		

Chapter 1 Numbers to One Million

In Dimensions Math 3, students learned to:

- Relate each digit of a four-digit number to its place value.
- Use place-value discs to show a four-digit number and locate it on a number line.
- Compare and order up to four-digit numbers.
- Round up to four-digit numbers to the nearest ten, hundred, and thousand.
- Create and interpret number patterns that increase or decrease by 1 ten, 1 hundred, or 1 thousand.

In Chapter 1, students will build on these concepts to extend knowledge and skills to five-digit and six-digit numbers. The chapter will help students understand how big a million is and understand the naming system. In the U.S. it is customary to separate periods of three digits with commas (for example, after the thousands and millions places). Other countries may use a dot or space to separate periods of numbers.

Place Value

The position of the digit in relation to other digits determines its value. Each place represents a value ten times the place to its right.

The number 865,423 is the same value as:

8 hundred thousands, 6 ten thousands, 5 thousands, 4 hundreds, 2 tens, 3 ones

In 865,423, the digit 6 is in the ten thousands place and stands for 60,000. Similarly, the value of the digit 4 in this number is 400.

The number 865,423 can also be written as a sum of its place values, which is called the expanded form of the number:

800,000 + 60,000 + 5,000 + 400 + 20 + 3

Place-value Discs

In order to gain a solid understanding of place value, students should have sufficient hands-on experience with manipulatives and see many different representations of place value.

Students should be familiar with base ten blocks from Dimensions Math 2 and 3. However, thousands, and larger numbers, are impractical to represent with base ten blocks.

In Dimensions Math 4 and 5, place-value discs will be used to introduce place value in decimal numbers.

Students have learned that:

- A 10-disc represents the value of 10 ones or 1 ten.
- A 100-disc represents the value of 10 tens or 1 hundred.
- A 1,000-disc represents the value of 10 hundreds or 1 thousand.

In this chapter, students will learn:

- A 10,000-disc represents the value of 10 thousands.
- A 100,000-disc represents the value of 10 ten thousands or 1 hundred thousand.

Place-value discs will be used to represent numbers periodically in the textbook when introducing new concepts.

Place-value Charts and Organizers

Students will continue to use a place-value chart, writing each digit of the number in the correct column according to its place:

Hundred Thousands	Ten Thousands	Thousands	Hundreds	Tens	Ones
8	6	5	4	2	3

If place-value discs are used, there are no headers indicating the value of each column:

Chapter 1 Numbers to One Million

Notes

Place-value Cards or Strips

Students should have their own sets of Place-value Cards (BLM) for numbers to the thousands. Place-value Cards (BLM) help students understand that numbers can be composed or decomposed according to the value of each digit.

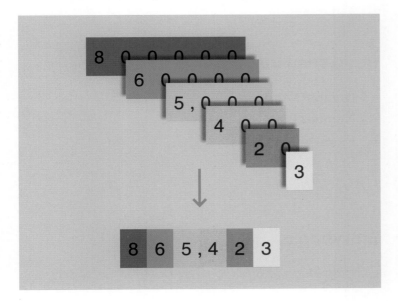

Number Lines and Rounding

Rounding is a skill that is used for estimating calculations. We also round in cases where it is more convenient or where it is difficult to accurately determine the real world measurement, such as the distance between two cities, the depth of a lake, or the height of a mountain.

This chapter formally introduces rounding five-digit and six-digit numbers. Number lines help students visualize the position of the number relative to the place-value to which they are rounding.

Students often have difficulty locating the position of a number on the number line. In the example of 25,446, rounding to the nearest thousand requires students to first identify the intervals between tick marks and then identify the nearest thousands (25,000 and 26,000).

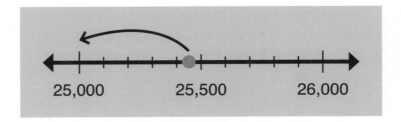

The number 25,446 rounded to the nearest thousand is 25,000.

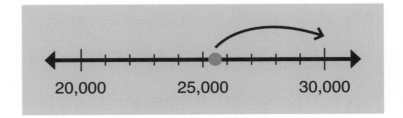

25,446 rounded to the nearest ten thousand is 30,000.

By convention, numbers that are halfway between, or equally far from the two multiples of ten, round to the greater number.

Students will learn which digits to focus on to round to a given place. For example, 25,466 could be rounded to 30,000, 25,000, or 25,500 depending on the prompt.

Students will use rules for rounding numbers to the nearest hundred, thousand, or ten thousand. The textbook states that the number is x when rounded to the nearest y with y being a multiple of 10, 100, or 1,000. For example, the number 21,646 rounded to the nearest thousand is 22,000.

Later in Chapter 1, students will learn that we can sometimes round to a number if that facilitates estimation. For example, to estimate the quotient of 21,646 ÷ 7, the nearest thousand is 22,000, which is not easy to divide by 7. Rounding 21,646 to 21,000 makes for a simpler calculation.

By working with number lines and rounding, students learn to determine numbers close to a given number as an estimate.

Chapter 1 Numbers to One Million

Calculations with Place Value

Students will continue to gain mastery of place value, and use that mastery to extend mental math strategies for up to two-digit numbers and larger numbers with up to 2 non-zero digits.

Examples:

140,000 + 30,000

Can be calculated by thinking:

- 14 ten thousands + 3 ten thousands = 17 ten thousands
- 140 thousands + 30 thousands = 170 thousands

Students can calculate these expressions using the same strategies they used with two-digit numbers (14 + 3 = 17).

120,000 − 30,000

Can be calculated by thinking:

- Since 12 − 3 = 9, 12 ten thousands − 3 ten thousands = 9 ten thousands, which we write as 90,000.

5,000 × 8

Some students rely on counting zeros to solve a problem such as 5,000 × 8. Because the basic fact 5 × 8 has a product ending in a zero, this can lead to confusion. Students can think of this as:

- 5 thousands × 8 = 40 thousands, which we write as 40,000.

420,000 ÷ 6

This can be calculated mentally by thinking of 42 divided by 6, remembering that it is 42 ten thousands, so the quotient will be a number of ten thousands:

- 42 ten thousands ÷ 6 = 7 ten thousands, which we write as 70,000.

Materials

- 10-sided die
- Bag
- Dry erase markers
- Dry erase sleeves
- Place-value discs
- Whiteboards

Blackline Masters

- Number Cards
- Number Line
- Place-value Cards
- Ship Weight
- Truck Weight

Storybooks

- *How Much is a Million?* by David M Schwartz

Activities

Games and activities included in this chapter provide practice and extensions of place-value concepts. They can be used after students complete the **Do** questions, or anytime review and practice are needed.

Chapter Opener

Objective

- Investigate large numbers.

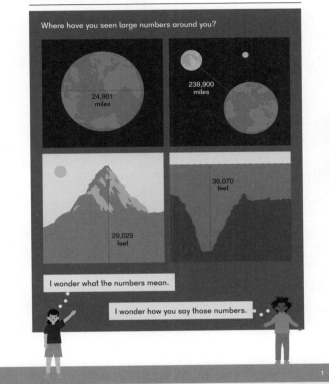

Have students discuss the **Chapter Opener**. They can look at the numbers to see if they can read them, and then talk about where they have seen or heard larger numbers.

Possible questions to ask students:

- "Have you heard the names of really large numbers?"
- "Is a zillion a number? Is a googol a number?"
- "What is the largest number you know?"

These are interesting questions to discuss and ponder. Answers are not required.

The book *How Much is a Million?* by David M Schwartz is also an engaging introduction to large numbers.

This introduction to large numbers can be a short discussion before beginning Lesson 1.

Lesson 1 Numbers to 100,000

Objectives

- Understand how numbers to 100,000 are composed based on place value.
- Write up to five-digit numbers using expanded form.

Lesson Materials

- Place-value Cards (BLM)
- Place-value discs to the thousands

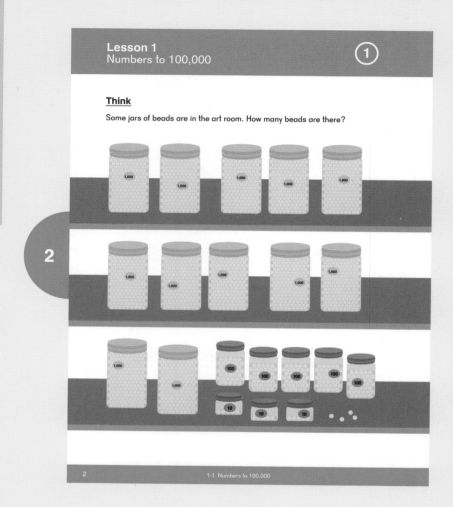

Think

Pose the **Think** problem and have pairs of students represent the total number of beads with place-value discs to the thousands only.

Discuss the strategies students used to find their answers to the **Think** problem.

Examples:

- "I counted 12,000, 500, 30, and 4."
- "I counted each value of place-value discs and found I had 12 thousands plus 5 hundreds plus 3 tens plus 4 ones."
- "I think I need a new place-value disc because I have more than 10 thousands."

Since students have only worked with four-digit numbers, they might point out that they need a new place-value disc because they have more than 10 thousands. Ask students what they think the new place value is called. If necessary, help them recall prior knowledge regarding the value of a digit on a place-value chart as it moves each place to the right. Introduce the ten thousand place and provide students with place-value discs for ten thousand.

Have students exchange 10 thousand discs for 1 ten thousand disc and show the number. They should see that 12 thousands is equal to 1 ten thousand and 2 thousands.

Learn

Students should see from the use of the place-value discs and cards that the written numeral shows the number of ten thousands, thousands, hundreds, tens, and ones from left to right.

Have students compare the different representations of the number 12,534:

- Place-value discs
- Place-value chart
- Place-value cards
- Number words
- Expanded form

Discuss Alex's and Sofia's comments about the groups of tens and hundreds in 10,000. Students will answer questions similar to these in ❼.

Alex's question, "How many groups of one hundred are in 10,000?" is not asking for the value of the digit in the hundreds place. He is asking, "How many hundred discs would have to be used to represent 10,000?"

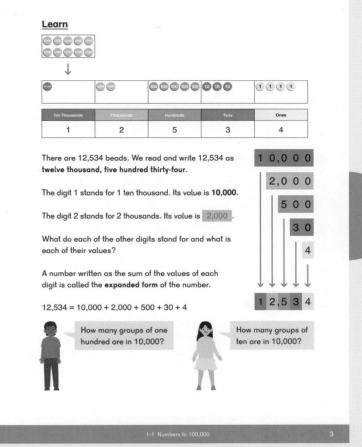

Ten Thousands	Thousands	Hundreds	Tens	Ones
1	2	5	3	4

10,000 is **100** hundreds.
10,000 is **1,000** tens.

Do

Have students use place-value discs if they need additional help understanding the value of the numbers.

1 Discuss how the given interval shows students how to label each tick mark.

3 Point out to students that any place with the value of 0 is not represented with a place-value card showing zeros only. In (a) the tens and in (b) the thousands are not represented with place-value cards showing zero.

4 Questions similar to those in (c) and (d) can be asked of any numbers to solidify understanding of place value:

- "What is the value of the digit _____ in each number?"
- "What does the digit _____ stand for in each number?"

5 – 7 Have students show the numbers on whiteboards to asses for understanding.

5 If needed, provide additional practice with different five-digit numbers and questions similar to (d) and (e), where the values given are not necessarily in order from greatest to least.

If students need additional help, have them use Place-value Cards (BLM) to make the numbers.

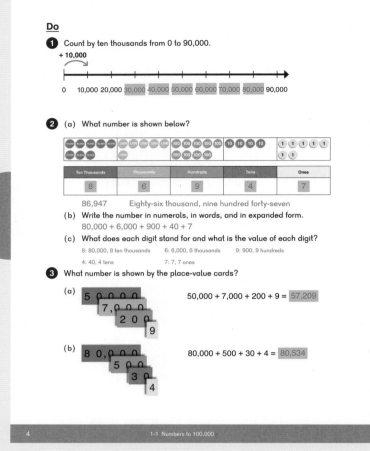

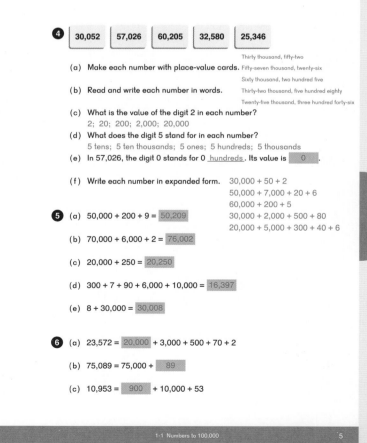

⑦ Ask students additional, similar questions to practice and solidify their understanding of place value.

For example:

- 34,000 = ▭ thousands
- 34,000 = ▭ hundreds
- 34,000 = ▭ tens
- 34,000 = ▭ ones

Activities

▲ Match Me

Materials: Place-value Cards (BLM)

One partner creates a five-digit number with the Place-value Cards (BLM). His partner writes the number in expanded form. If a third student is playing, she can write the number in words.

Students could also create the same number using place-value discs, but not in the same way shown by the other players. For example, for 31,425, instead of showing 3 ten thousand discs and 1 thousand disc, the player could show 3 ten thousand discs, 14 hundred discs, and 25 ones discs.

▲ Grab and Build

Materials: Place-value discs in a bag with additional discs for regrouping

Students can grab a handful of discs from the bag and organize them into a number. They should regroup ten of any place in their handful for the next greatest place value. Once they have created their number, have them record it in standard form, expanded form, and word form.

For example:

21,264
20,000 + 1,000 + 200 + 60 + 4
Twenty-one thousand, two hundred sixty-four

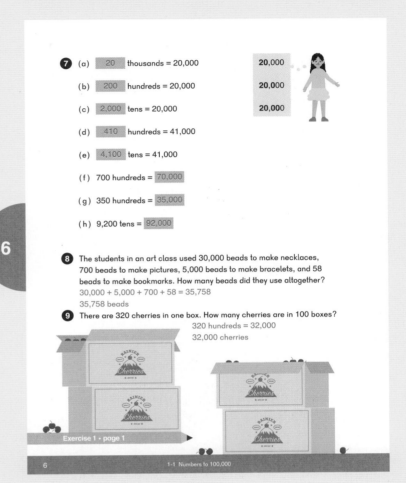

Lesson 2 Numbers to 1,000,000

Objectives

- Understand how numbers to 1,000,000 are composed based on place value.
- Write up to six-digit numbers using expanded form.

Lesson Materials

- Place-value Cards (BLM)
- Place-value discs to the hundred thousands

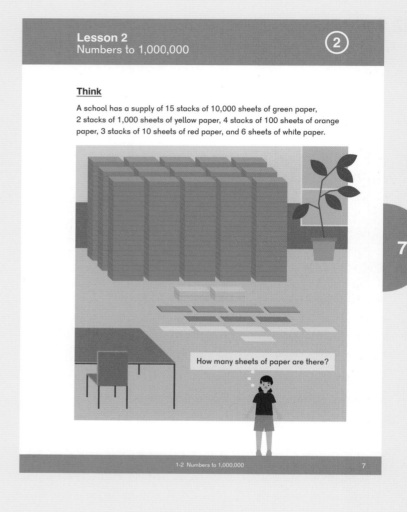

Think

Pose the **Think** problem and have pairs of students represent the number of stacks of paper with Place-value Cards (BLM) or place-value discs.

Discuss the strategies students used to find their answers to the **Think** problem.

As in the previous lesson, ask students what they think the new place value is called and introduce the new place value: hundred thousands. Then provide place-value discs for hundred thousands.

Have students exchange 10 ten thousand discs for 1 hundred thousand disc and represent the number in the problem using this new place-value disc. They should see that 15 ten thousands is equal to 1 hundred thousand and 5 ten thousands.

Learn

As in the previous lesson, students should see from the use of the place-value cards that the written numeral shows the number of hundred thousands, ten thousands, thousands, hundreds, tens, and ones from left to right.

Have students discuss the different representations of the number 152,436:

- Place-value discs
- Place-value cards
- Place-value chart
- Number words
- Expanded form

Discuss Mei's and Emma's questions about the number of thousands and hundreds in 100,000. Students will answer additional questions similar to these in ❼.

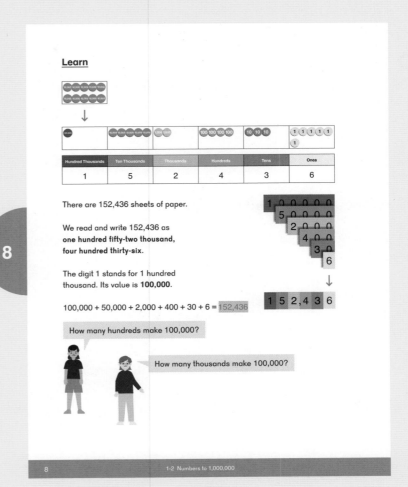

Do

1 Ten hundred thousands shows the need for another place as 1 million is introduced. Tell students that regrouping 10 hundred thousands as 1 million creates a new place to the left of the hundred thousands place.

Point out the placement of the comma between the millions and hundred thousands places.

Students should recall that a digit to the left of the same digit on a place-value chart is ten times the value of the digit to the right.

Ask students:

- "What does that tell us about a digit that is one place to the left of a digit in the hundred thousands place?"
- "Ten hundred thousands is equal to what number?"
- "Do you think this pattern of regrouping ten of one unit and exchanging them for one of the next greater place continues?"
- "What might the next greater place be?"

5 As an extension to (c) and (d), ask additional questions using six-digit numbers:

- "What is the value of the digit _____ in each number?"
- "What does the digit _____ stand for in each number?"

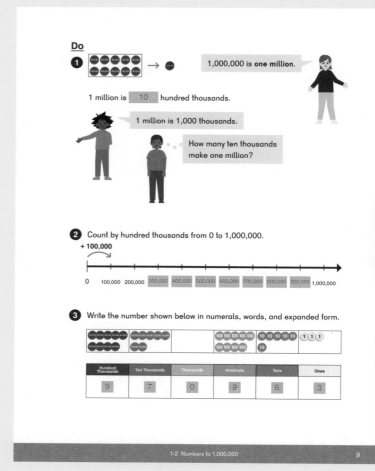

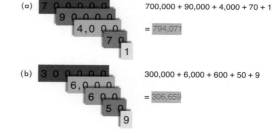

6 If students need additional help, provide them with Place-value Cards (BLM) to make the number.

7 Ask students additional questions to practice and solidify their understanding of place value. These skills will be necessary for Lesson 3.

For example:

- 340,000 = ☐ ten thousands
- 340,000 = ☐ thousands
- 340,000 = ☐ hundreds
- 340,000 = ☐ tens

▲ Activity

Place Value 20 Questions

Students play **Place Value 20 Questions** using six-digit numbers. Player One chooses a six-digit number in which no digits repeat. He draws 6 lines:

___ ___ ___ ___ ___ ___

Player Two guesses with questions such as:

- "Is the ten thousands digit greater or less than 4?"
- "Is the digit in the tens place greater than 5?"
- "Is the value of the hundreds digit less than 400?"
- "Is the digit in the hundred thousands place odd?"

Exercise 2 · page 4

6
(a) 500,000 + 60,000 + 4,000 + 200 + 7 = 564,207
(b) 200 + 3 + 400,000 + 6,000 + 50,000 + 90 = 456,293
(c) 600,000 + 20,000 = 620,000
(d) 1,000 + 30 + 40,000 = 41,030
(e) 351,200 = 350,000 + 1,200
(f) 80 + 300,000 + 7,000 = 307,080

7
(a) 20 ten thousands = 200,000
(b) 4,000 hundreds = 400,000
(c) 97,000 tens = 970,000
(d) 100 ten thousands = 1,000,000

8
A school ordered 100,000 red streamers and 70,000 green streamers. To thank the school for the large order, the company also sent 200 silver streamers as a gift. How many streamers did the school receive in all?
100,000 + 70,000 + 200 = 170,200
170,200 streamers

9 How many sheets of paper are in one thousand packs of 750 sheets?
750 thousands = 750,000
750,000 sheets of paper

Exercise 2 · page 4

Lesson 3 Number Patterns

Objectives

- Find a number that is 100, 1,000, or 10,000 more or less than a five-digit number.
- Count on and back from a given number by hundreds, thousands, or ten thousands.

Lesson Materials

- Place-value Cards (BLM)
- Place-value discs

Think

Provide students with place-value discs and have them solve the **Think** problems.

Discuss student strategies for solving the problems.

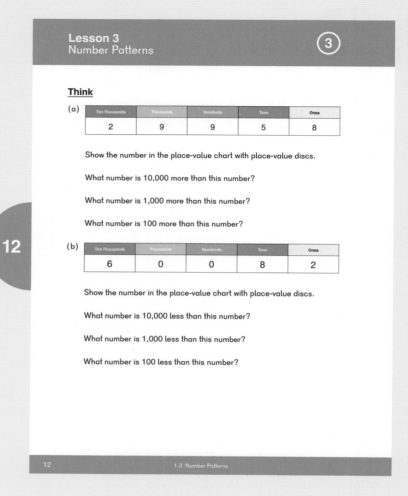

Learn

Note that the emphasis is on the place value of the digits in the specific numbers in the equations. Students may have more difficulty with the questions asking what is 1,000 or 100 more or less than a number.

Discuss how the place value language that Mei, Sofia, and Alex use in **Learn** makes the questions easier to answer with mental math. For example, 2 ten thousands + 1 ten thousand = ? ten thousands.

(a) Mei prompts students to add 1 ten thousand disc.

Sofia points out that adding 1 thousand disc requires regrouping to the next greatest place value, ten thousand.

Alex points out that adding 1 hundred disc requires regrouping to the next greatest place value, thousand. Regrouping to the thousands place requires further regrouping to the next greatest place value, ten thousand.

(b) Discuss the similar situations for the three subtraction problems.

Ask students which place values they must be concerned with to answer each of the questions. Students do not need to consider the tens or ones places in any of the questions.

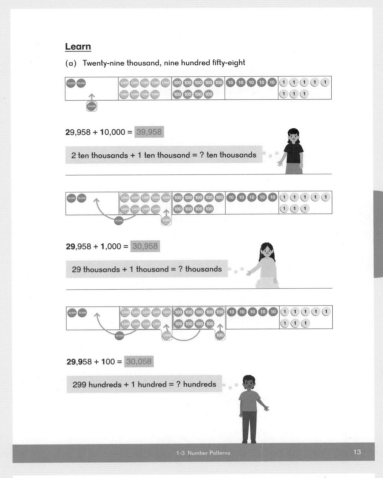

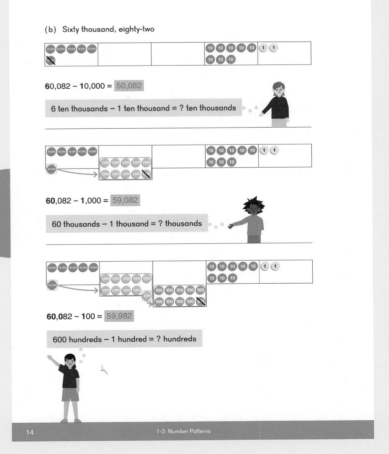

Do

1–**2** Have students work in pairs with Place-value Cards (BLM). In each problem, students should think about the place value that is changing.

Examples:

1 (b) 63 thousands + 1 thousand = 64 thousands, so 64,547.

(c) 635 hundreds + 1 hundred = 636 hundreds, so 63,647.

2 (b) 49 thousands − 1 thousand = 48 thousands, so 48,386.

(c) 493 hundreds − 1 hundred = 492 hundreds, so 49,286.

3–**4** Have students work in pairs with place-value discs to solve the problems.

Discuss Sofia's question to point out when it is necessary to regroup in each place.

5–**6** Students should be able to complete the problems without place-value discs.

In (a) students could think: 53, 43, 33, 23, 13, 3 thousands. The other values in the number do not change.

Activity

▲ **What's My Rule?**

Students take turns being the Ruler and the Solver. The Ruler writes a five-digit number and a rule for generating a sequence of numbers using 10,000, 1,000, or 100.

The Solver writes the next five numbers that follow the rule.

For example:

The Ruler writes, "35,687" and "+ 1,000."

The Solver needs to find: 36,687, 37,687, 38,687, 39,687, and 40,687.

★ Challenge students with a two-step rule.

For example:

The Ruler writes, "47,447" and "+ 1,000 – 100."

The Solver needs to find: 48, 347, 49,247, 50,147, 51,047, and 51,947.

Exercise 3 • page 7

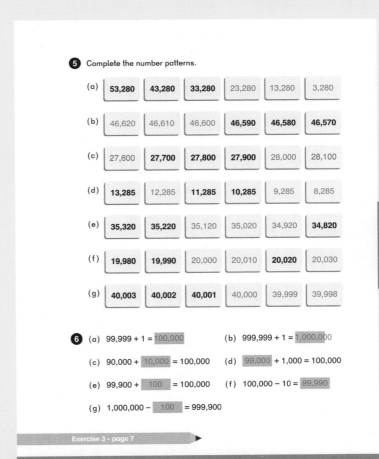

Lesson 4 Comparing and Ordering Numbers

Objective

- Compare and order numbers within 1 million.

Lesson Materials

- Place-value Cards (BLM)

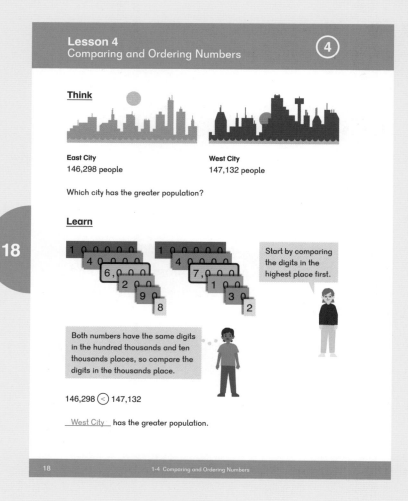

Think

Provide students with Place-value Cards (BLM) and pose the **Think** question.

Ask what strategies they could use to determine which city has the greater population.

Discuss the different strategies.

Examples:

- "147 thousand is greater than 146 thousand."
- "I just compared the digits in each place value to see which was greater."
- "6 thousands is less than 7 thousands so West City has the greater population."

Learn

Discuss Emma's comment. Have students compare their solutions from **Think** with the one shown in the textbook.

Guide students to look at the digits in the place values from greatest to least, left to right.

Review the meaning of the "greater than" and "less than" signs.

Do

1 — 4 Discuss Emma's, Alex's, and Mei's thoughts.

3 — 4 Mei and Dion get students to see that they can line the numbers up and compare the digits in each place. Mei shows students that they should start from the greatest place value.

5 Encourage students who need additional help to line the numbers up on a place-value chart or make the numbers with Place-value Cards (BLM). This will help them compare the place values.

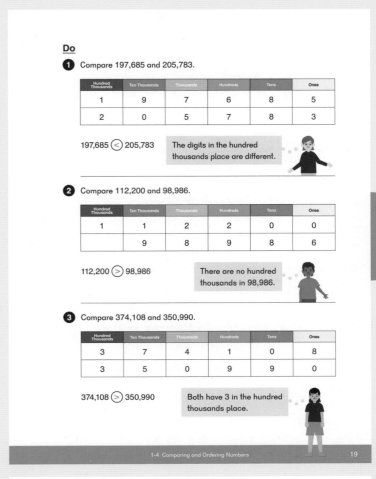

❻ This problem could be turned into an activity by writing the numbers on index cards and having students line the cards up in increasing order.

Activity

▲ Greatest or Least?

Materials: Number Cards (BLM) 0—9 or 10-sided die, whiteboards, dry erase markers

Play with entire class:

Players draw a six-digit place-value chart and one extra box labeled "trash" as shown below.

The teacher draws a number card and shows it to the students. Each player decides in which place to write the number before the next card is drawn by the teacher.

Hundred Thousands	Ten Thousands	Thousands	Hundreds	Tens	Ones

Trash

Once the digit has been written, it must stay in that place.

Players may write one digit in the box labeled "trash." When 7 cards have been drawn, the players with the greatest numbers are the winners.

Modify to play for the least six-digit number.

Play in groups of up to 4 players:

Instead of the teacher drawing cards, players take turns drawing number cards and writing each digit in different places on their game boards.

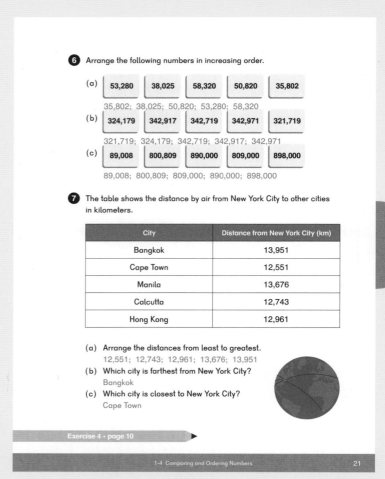

After all players have drawn 7 number cards, they compare their numbers. The player with the greatest six-digit number wins.

Exercise 4 • page 10

Lesson 5 Rounding 5-digit Numbers

Objective

- Round a five-digit number to the nearest ten thousand, thousand, or hundred.

Lesson Materials

- Dry erase sleeves
- Number Line (BLM)
- Truck Weight (BLM)

Think

Pose the **Think** problem and ask students to recall and share what they remember about rounding numbers.

Ask students:

- "What numbers are the mountain heights between?"
- "What are the nearest 10,000 ft to each mountain height? One increment should be less than and one increment should be greater than each mountain height."
- "What numbers do the tick marks denote?"
- "What are the increments between the tick marks?"

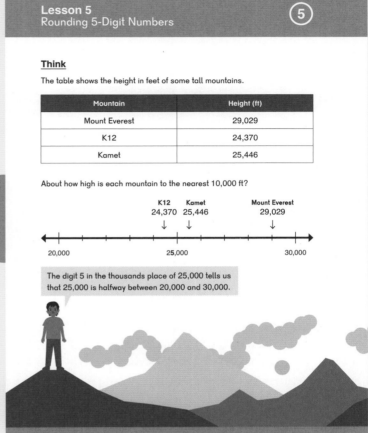

Learn

Discuss how students would place the numbers on the number line to round to the nearest 10,000. 20,000 is the nearest ten thousand to 24,370 and 30,000 is the nearest ten thousand to 29,029.

To place the heights from **Think** on Number Line (BLM), students would label the tick marks for 20,000 and 30,000, then count on by 1,000 from one number to the other.

Because the numbers are so large, it is hard to label every tick mark on the number line. To help students see half, Alex (in **Think**) points out the labeled halfway mark between 20,000 and 30,000.

Discuss the textbook explanations.

Since the numbers are so large, students will need to estimate where the heights belong on the number line. They should see that it is not possible to mark 29,029 on the number line exactly.

Mei reminds us of the convention that when we are halfway between ten thousands, or equally far from both benchmarks, we round to the greater benchmark.

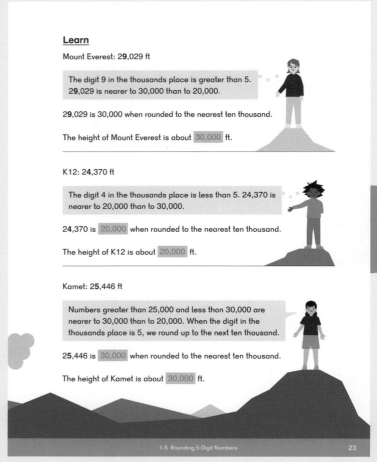

Do

① (b) When rounding to the nearest thousand, students first need to identify the thousands that are nearest to 62,952 (62,000 and 63,000).

(c) When rounding to the nearest hundred, students first need to identify the hundreds that are nearest to 62,952 (62,900 and 63,000).

Mei reminds students to round 52 up to the next hundred.

⑤ Provide students with Truck Weight (BLM) if needed.

Activity

▲ **Round to 30,000**

Materials: Number Cards (BLM) 1–6

Using any five of the six digits 1 through 6, one time each, find the largest number that rounds to 30,000.

Find the smallest number that rounds to 30,000.

★ Extend the activity by providing additional rules. For example, ask students: "Find the smallest odd number that rounds to 30,000 in which the value of the digit in the ones place exceeds the value of the digit in the hundreds place."

Exercise 5 • page 13

Lesson 6 Rounding 6-digit Numbers

Objective

- Round a six-digit number to the nearest hundred thousand, ten thousand, or thousand.

Lesson Materials

- Dry erase sleeves
- Number Line (BLM)
- Ship Weight (BLM)

Think

Pose the **Think** problem and provide adequate time for students to find solutions. Ask students to share how they rounded the numbers.

Learn

As in Lesson 5, discuss how the numbers are placed on the number line when rounding to the nearest hundred thousand. 600,000 is the nearest hundred thousand less than 631,346, and 700,000 is the nearest hundred thousand greater than 681,124.

When placing the city populations from **Think** on Number Line (BLM), students should see that the value of the intervals between tick marks is 10,000.

Just as with five-digit numbers, it is not necessary to label every tick mark on the number line. To help students estimate and find the midpoint, Alex points out the labeled halfway mark between 600,000 and 700,000.

Discuss the textbook explanations.

Sofia reminds us that when we are halfway between two increments, we round to the greater number.

Lesson 6
Rounding 6-Digit Numbers

Think

This table shows the population of some cities.

City	Population (Number of people)
Oklahoma City	631,346
El Paso	681,124
Memphis	655,770

What is the population of each city to the nearest hundred thousand?

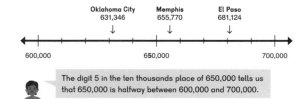

The digit 5 in the ten thousands place of 650,000 tells us that 650,000 is halfway between 600,000 and 700,000.

26　　　　1-6 Rounding 6-Digit Numbers

Learn

Oklahoma City: 631,346

The digit 3 in the ten thousands place is less than 5.

631,346 is 600,000 when rounded to the nearest hundred thousand.

The population of Oklahoma City is about 600,000 people.

El Paso: 681,124

The digit 8 in the ten thousands place is greater than 5.

681,124 is 700,000 when rounded to the nearest hundred thousand.

The population of El Paso is about 700,000 people.

Memphis: 655,770

The digit in the ten thousands place is 5. We round up to the next hundred thousand.

655,770 is 700,000 when rounded to the nearest hundred thousand.

The population of Memphis is about 700,000 people.

Teacher's Guide 4A Chapter 1　　© 2019 Singapore Math Inc.

Do

1 (b) When rounding to the nearest ten thousand, students first need to identify the ten thousands that are nearest to 748,500 (740,000 and 750,000).

(c) When rounding to the nearest thousand, students first need to identify the thousands that are nearest to 748,500 (748,000 and 749,000).

Mei reminds students to round 500 up to the next thousand.

5 Provide students with Ship Weight (BLM) if necessary.

Activity

▲ Round to Hundred Thousands

Materials: Number Cards (BLM) 1–6
Provide students with number cards and have them make 6-digit numbers that when rounded to the nearest hundred thousand would be:

- 700,000
- 600,000
- 500,000
- 400,000
- 300,000
- 200,000
- 100,000

★ Extend the activity by providing additional rules.

For example, ask students:

"Find an odd number that rounds to 600,000 in which the value of the digit in the tens place exceeds the value of digit in the hundreds place."

Exercise 6 • page 16

Do

1 (a) Round 748,500 to the nearest hundred thousand.

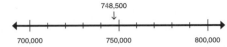

748,500 is 700,000 when rounded to the nearest hundred thousand.

(b) Round 748,500 to the nearest ten thousand.

748,500 is 750,000 when rounded to the nearest ten thousand.

(c) Round 748,500 to the nearest thousand.

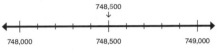

 The digit 5 is in the hundreds place. We round up to the next thousand.

748,500 is 749,000 when rounded to the nearest thousand.

2 Round each number to the nearest hundred thousand.

(a) 485,000   (b) 130,498   (c) 150,873
    500,000   100,000   200,000

3 Round each number to the nearest ten thousand.

(a) 285,000   (b) 160,498   (c) 827,055
    290,000   160,000   830,000

4 Round each number to the nearest thousand.

(a) 640,989   (b) 345,500   (c) 297,368
    641,000   346,000   297,000

5 The table shows the weight in tons of some different types of cruise ships. Copy and complete the table.

Ship	Weight (tons)	Weight to the nearest 100,000 tons	Weight to the nearest 10,000 tons	Weight to the nearest 1,000 tons
Oasis of the Seas	225,282	200,000	230,000	225,000
Ovation of the Seas	168,666	200,000	170,000	169,000
Queen Mary 2	149,215	100,000	150,000	149,000
Navigator of the Seas	139,570	100,000	140,000	140,000
Celebrity Eclipse	121,878	100,000	120,000	122,000

A ton is a unit of weight equal to 2,000 pounds. We use tons to express the weight of heavy objects.

Exercise 6 • page 16

Lesson 7 Calculations and Place Value

Objective

- Use place value concepts in mental computations.

Lesson Materials

- Place-value discs

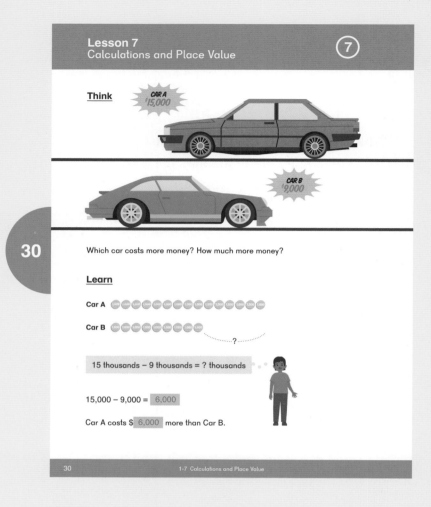

Think

Provide students with place-value discs and have them solve the **Think** questions. Discuss student strategies for solving the problem.

Learn

Help students see that they can use mental math to easily calculate large numbers in which most of the digits are 0. Alex demonstrates this thinking. Students see that solving 15,000 − 9,000 is just like solving the simpler computation, 15 − 9, but rather than the units being ones, i.e., 15 ones − 9 ones = 6 ones, the units are thousands.

Students should use mental math, not the vertical algorithm, to solve the problems in this lesson.

Do

1–**3** The friends demonstrate how to find the answer by using a simpler computation with thousands or ten thousands as the unit.

Discuss each problem and how the friends apply the methods of this unit to addition, multiplication, and division.

1 If thousands are the units, students can think of the problem as 170 thousands + 80 thousands = 250 thousands.

5 Students should think of ten thousands as the unit for the product, or 4 ten thousands, which is written 40,000. This language will help students understand the units, not merely "count zeros."

6 20 hundreds ÷ 5 is an easier computation than 2 thousands ÷ 5.

7 (g) 40 ten thousands ÷ 8 is an easier computation than 4 hundred thousands ÷ 8.

Exercise 7 • page 19

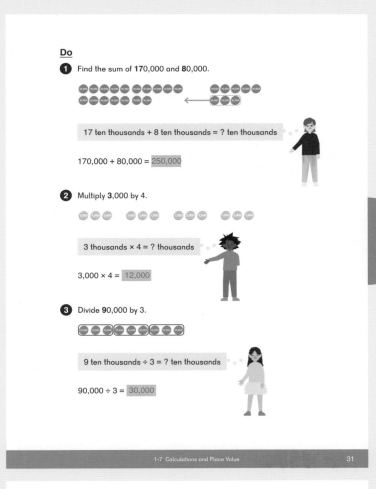

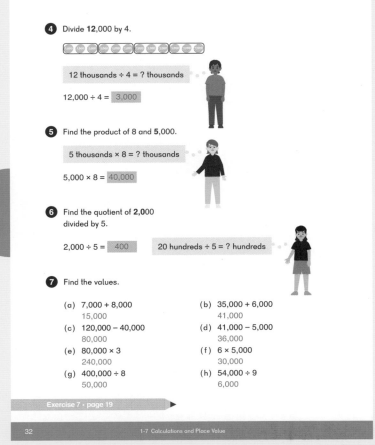

Lesson 8 Practice

Objective

- Practice concepts from the chapter.

After students complete **Practice** in the textbook, have them continue working with place value and rounding by playing games from the chapter.

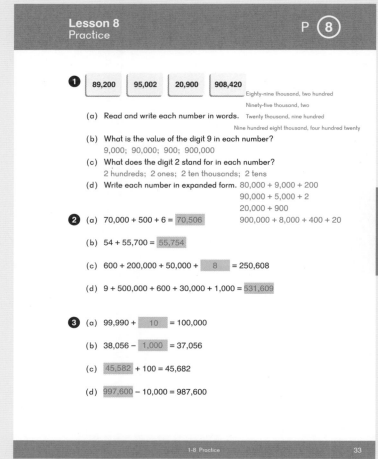

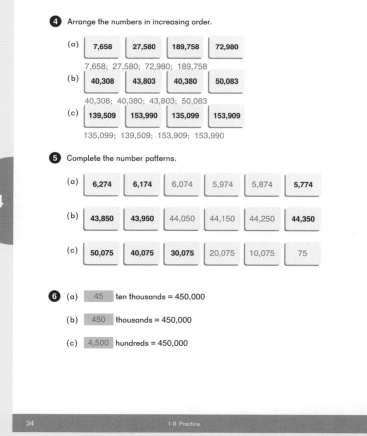

14 If students mistakenly multiply, ask:

- "What do we know?"
- "What are we trying to find?"

Have them draw a model:

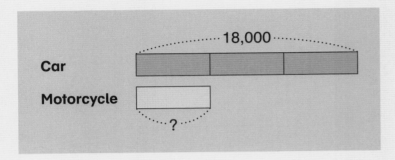

3 units ⟶ 18,000
1 unit ⟶ 18,000 ÷ 3 = 6,000

Exercise 8 • page 22

Brain Works

★ What Number Am I? Riddles

- I am a five-digit number.
- When rounded to the nearest thousand, I am 46,000.
- My digits in the thousands, hundreds, and tens places are all different odd numbers.
- My ten thousands digit is one less than my thousands digit.
- My hundreds digit is less than 9.
- My tens and ones digits form a number divisible by both 6 and 8.
- What number am I? 45,796

- I am a five-digit number with no digits that repeat.
- The sum of the digit in the ten thousands place and the ones place is 8 and their product is 0.
- The digit in the tens place is twice the digit in the thousands place. Their product is 18.
- The digit in the ten thousands place is twice the digit in the hundreds place.
- What number am I? 83,460

Challenge students to write their own riddles.

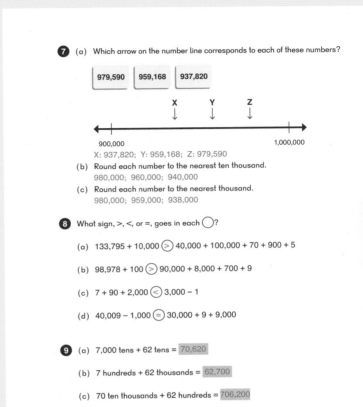

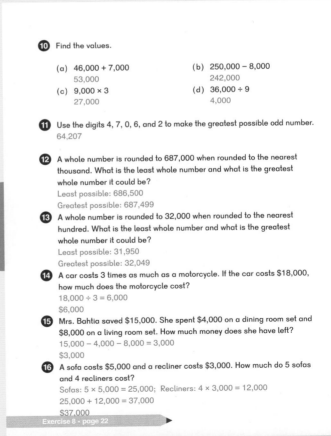

Exercise 1 • pages 1–3

Chapter 1 Numbers to One Million

Exercise 1

Basics

1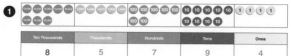

Ten Thousands	Thousands	Hundreds	Tens	Ones
8	5	7	9	4

(a) Complete the place-value chart.

(b) Write the number in numerals. 85,794

(c) The value of the digit in the ten thousands place is __80,000__.

(d) The digit 5 in this number stands for 5 __thousands__.

(e) Write the number in expanded form. 80,000 + 5,000 + 700 + 90 + 4

Practice

2 Write the value of each bolded digit.

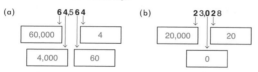

(c) In 64,564, the digit 5 stands for 5 __hundreds__.

(d) In 23,028, the digit __3__ is in the thousands place.

3 (a) __87,143__ = 80,000 + 7,000 + 100 + 40 + 3

(b) 60,437 = 60,000 + __400__ + 30 + 7

(c) 14,092 = 10,000 + __4,000__ + 92

(d) __41,404__ = 400 + 1,000 + 40,000 + 4

(e) 90,620 = 600 + __90,000__ + 20

(f) 30,030 = __30__ + 30,000

(g) 62,500 = __500__ + 2,000 + 60,000

4 Write the number in numerals.

forty-seven thousand, six hundred ninety-eight	47,698
twenty-three thousand, two hundred four	23,204
thirty-three thousand, thirty-one	33,031
eighteen thousand, forty	18,040
eighty thousand, nine	80,009

5 Write the number in words.

20,640	twenty thousand, six hundred forty
98,700	ninety-eight thousand, seven hundred
55,008	fifty-five thousand, eight
90,077	ninety thousand, seventy-seven
12,120	twelve thousand, one hundred twenty

6 (a) 60,000 = __6__ ten thousands

(b) 60,000 = __60__ thousands

(c) 60,000 = __600__ hundreds

(d) 60,000 = __6,000__ tens

(e) 35,000 = __350__ hundreds

(f) 10,100 = __1,010__ tens

7 Write the number in numerals.

723 hundreds	72,300
9 ten thousands + 24 tens	90,240
800 hundreds + 60 hundreds + 50 tens	86,500
640 tens + 20 thousands	26,400
92 ones + 4,800 tens	48,092

Challenge

8 Use the clues to find the mystery 5-digit number.

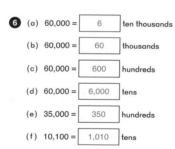

Use trial and error with 7, 8, 9 in thousands place. 9 and 8 will cause repeated digits. So it could be __7 8 0 1 or __7 9 0 2.

Clue 1 All the digits are different.
Clue 2 The thousands digit is 7 more than the tens digit.
Clue 3 The hundreds digit is the sum of the thousands digit and the ones digit.
Clue 4 The total of the digits when added together is 24.

7 + 8 + 0 + 1 = 16; 24 − 16 = 8; repeated digit
The number is __67,902__.   7 + 9 + 0 + 2 = 18; 24 − 18 = 6

Exercise 2 • pages 4–6

Exercise 2

Basics

1.

Hundred Thousands	Ten Thousands	Thousands	Hundreds	Tens	Ones
6	9	8	3	6	5

(a) Write the number shown in numerals. 698,365

(b) The value of the digit in the hundred thousands place is 600,000 .

(c) The digit 3 in this number stands for 3 hundreds .

(d) Write the number in words.

six hundred ninety-eight thousand, three hundred sixty-five

(e) Write the number in expanded form.

600,000 + 90,000 + 8,000 + 300 + 60 + 5

Practice

2. Write the value of each bolded digit.

(a) **3**8**4**80**2**
- 300,000
- 0
- 80,000
- 800

(b) **4**9**4**4**2**8
- 400,000
- 400
- 4,000

(c) In 384,802, the digit 4 stands for 4 thousands .

(d) In 494,428, the digit 9 is in the ten thousands place.

3. (a) 625,239 = 600,000 + 20,000 + 5,000 + 200 + 30 + 9

(b) 160,330 = 100,000 + 60,000 + 300 + 30

(c) 604,085 = 600,000 + 4,000 + 80 + 5

(d) 914,053 = 4,000 + 900,000 + 3 + 10,000 + 50

(e) 110,680 = 600 + 10,000 + 100,000 + 80

(f) 720,076 = 6 + 70 + 20,000 + 700,000

(g) 500,200 = 200 + 500,000

4. Write the number in numerals.

four hundred thousand, six hundred ninety-eight	400,698
seven hundred twenty-three thousand, one	723,001
eight hundred thousand, forty	800,040
one hundred thirty thousand, thirty-one	130,031
one million	1,000,000

5. Write the number in words.

271,644	two hundred seventy-one thousand, six hundred forty-four
110,990	one hundred ten thousand, nine hundred ninety
199,009	one hundred ninety-nine thousand, nine
100,007	one hundred thousand, seven

6. (a) 200,000 = 20 ten thousands

(b) 200,000 = 2,000 hundreds

(c) 120,000 = 12,000 tens

(d) 320,000 = 320 thousands

7. Write the number in numerals.

9,860 hundreds	986,000
80 ten thousands + 60 hundreds + 5 tens	806,050
6,175 tens + 200 thousands	261,750
98 tens + 48 ten thousands	480,980
8 ones + 12,000 tens	120,008
ten hundred thousands	1,000,000

Challenge

8. Use the clues to find the mystery 6-digit number.
Answers may vary.
Clue 1 Only two of the digits are the same.
Clue 2 The digit in the tens place is 9 more than the digit in the thousands place. __ __ 0 __ 9 __
Clue 3 The number is an odd number.
Clue 4 The digit in the ones place is 5 more than the digit in the hundred thousands place. 2 __ __ 0 __ 9 7
Clue 5 The sum of the digits is 34.
2 + 0 + 9 + 7 = 18; 34 − 18 = 16; The other two digits are 8

The number is 280,897 .

Exercise 3 • pages 7–9

Exercise 3

Basics

1. Draw more discs or cross off discs to show the number, and fill in the blanks.

   1,000 more than 62,584 is __63,584__.

   100 more than 58,927 is __59,027__.

   10,000 less than 86,584 is __76,584__.

   10 less than 26,009 is __25,999__.

2. (a) 322,523 + 10,000 = 332,523 (b) 141,523 − 10 = 141,513
   (c) 14,690 + 100,000 = 114,690 (d) 81,096 − 100 = 80,996
   (e) 100,047 − 100 = 99,947 (f) 179,992 + 10 = 180,002

Practice

3. Follow the rules to create the number patterns.

   (a) Count on by ten thousand.

   | 97,376 | 107,376 | 117,376 | 127,376 | 137,376 |

   (b) Count back by one hundred.

   | 333,333 | 333,233 | 333,133 | 333,033 | 332,933 |

   (c) Count on by one thousand.

   | 208,997 | 209,997 | 210,997 | 211,997 | 212,997 |

4. Complete the number patterns.

68,821	68,831	68,841	68,851	68,861	68,871
		68,941			78,871
69,039	69,040	69,041	69,042		88,871
		69,141			98,871
438,876			141,873		108,871
338,876			140,873		118,871
238,876			139,873		128,871
138,876	138,875	138,874	138,873	138,872	138,871

5. (a) 689,423 is 1,000 more than __688,423__.
   (b) 111,111 is 10,000 more than __101,111__.
   (c) __40,989__ is 10,000 less than 50,989.
   (d) 400,912 is 10,000 more than __390,912__.
   (e) __98,999__ is 1,000 less than 99,999.
   (f) __100,099__ is 100 more than 99,999.
   (g) __1,000,000__ is 1 more than 999,999.

6. (a) 24,608 + __100__ = 24,708
   (b) 7,012 − __100__ = 6,912
   (c) 1,091 − __10__ = 1,081
   (d) 8,219 − __100__ = 8,119
   (e) 8,930 + __100__ = 9,030

Challenge

7. Complete the number patterns.
   1,000 more and 1 less
   (a) | **187,174** | **188,173** | **189,172** | 190,171 | 191,170 | 192,169 |
   (b) | **51,146** | **62,136** | **73,126** | 84,116 | 95,106 | 106,096 |
   10,000 more, 1,000 more, 10 less

Exercise 4 • pages 10–12

Exercise 4

Basics

1 Compare the numbers.

Hundred Thousands	Ten Thousands	Thousands	Hundreds	Tens	Ones
6	3	6	7	9	8
6	3	6	9	2	4

__636,924__ is greater than __636,798__.

Hundred Thousands	Ten Thousands	Thousands	Hundreds	Tens	Ones
4	9	8	3	6	5
4	2	8	9	0	0

__498,365__ is greater than __428,900__.

Hundred Thousands	Ten Thousands	Thousands	Hundreds	Tens	Ones
2	8	0	7	8	2
2	8	0	7	4	6

__280,746__ is less than __280,782__.

Hundred Thousands	Ten Thousands	Thousands	Hundreds	Tens	Ones
	9	8	3	6	5
1	7	3	0	0	8

__98,365__ is less than __173,008__.

Practice

2 Locate the numbers on the number line, then write them in order from least to greatest.

| 45,800 | 56,500 | 65,500 | 36,500 | 56,600 |

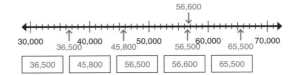

| 36,500 | 45,800 | 56,500 | 56,600 | 65,500 |

3 Write > or < in each ◯.

(a) 82,262 < 82,623 (b) 918,532 < 981,352

(c) 672,365 < 672,369 (d) 99,856 < 99,865

4 Cross out the numbers that are greater than 55,000 and circle the numbers that are less than 54,600.

| 45,993 | 54,393 | 7,713 | ~~55,739~~ | ~~315,193~~ |

5 Write the numbers in order from least to greatest.

| 792,793 | 797,913 | 983,713 | 792,739 | 797,193 |

| 792,739 | 792,793 | 797,193 | 797,913 | 983,713 |

6 (a) Use the digits 2, 4, 9, 6, 8, and 7 to form the numbers. Use each digit only once for each number.

The greatest 6-digit number.	987,642
The least 6-digit number.	246,789
The least 6-digit even number.	246,798
The greatest 6-digit odd number.	986,427

(b) Write the numbers in order from greatest to least.

| 987,642 | 986,427 | 246,798 | 246,789 |

7 Write > or < in each ◯.

(a) 34,009 + 40 > 8,000 + 4 + 20,000

(b) 90,000 + 60 + 4,200 < 300 + 94,000 + 20

(c) 300 thousands + 70 hundreds < 7,000 tens + 30 ten thousands

8 Read the clues. Then circle the correct number.

Clue 1 The sum of the digits is more than 20.
Clue 2 There are more than 2,000 tens.
Clue 3 There are less than 450 hundreds.

| 38,041 | 26,903 | 15,987 | 40,674 | 135,643 |

Exercise 5 • pages 13–15

Exercise 5

Basics

1. Fill in the blanks.

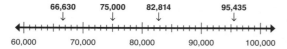

   (a) 66,630 is between 60,000 and 70,000. It is nearer to 70,000 than to 60,000. 66,630 is 70,000 when rounded to the nearest __ten thousand__.

   (b) 75,000 is halfway between 70,000 and __80,000__. 75,000 is __80,000__ when rounded to the nearest ten thousand.

   (c) 82,814 is between 80,000 and 90,000. It is nearer to __80,000__ than to __90,000__. 82,814 is __80,000__ when rounded to the nearest ten thousand.

   (d) 95,435 is __100,000__ when rounded to the nearest ten thousand.

   (e) When rounding to the nearest ten thousand, if the digit in the __thousands__ place is __5__ or more we round up to the next ten thousand. When it is __4__ or less we round down.

2. Round 48,665 to the nearest thousand.

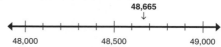

   48,665 is __49,000__ when rounded to the nearest thousand.

3. Round 10,025 to the nearest hundred.

   10,025 is __10,000__ when rounded to the nearest hundred.

Practice

4. Indicate the location or approximate location of each number on the number line. Then round each number to the nearest ten thousand.

   A 65,501 __70,000__   B 75,000 __80,000__
   C 51,980 __50,000__   D 64,999 __60,000__

5. Jupiter has a diameter of 88,846 miles at its equator. Round this number to the nearest ten thousand.   __90,000__

6. Round each number to the nearest ten thousand.

   (a) 10,920 __10,000__   (b) 16,501 __20,000__
   (c) 24,499 __20,000__   (d) 97,522 __100,000__

7. The table shows the maximum depth of some ocean trenches in feet. Complete the table.

Trench	Depth (ft)	Depth to the nearest		
		10,000 ft	1,000 ft	100 ft
Peru-Chile Trench	26,460	30,000	26,000	26,500
Kermadec Trench	32,962	30,000	33,000	33,000
Japan Trench	34,039	30,000	34,000	34,000
Tonga Trench	35,702	40,000	36,000	35,700
Mariana Trench	36,070	40,000	36,000	36,100

8. (a) What is the least whole number that rounds to 230,000 when rounded to the nearest ten thousand?   __225,000__

   (b) What is the greatest whole number that rounds to 230,000 when rounded to the nearest ten thousand?   __234,999__

Exercise 6 • pages 16–18

Exercise 6

Basics

1. Fill in the blanks.

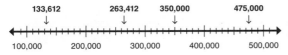

   (a) 133,612 is between 100,000 and 200,000. It is nearer to 100,000 than to 200,000. 133,612 is 100,000 when rounded to the nearest __hundred thousand__.

   (b) 263,412 is between 200,000 and 300,000. It is nearer to 300,000 than to 200,000. 263,412 is __300,000__ when rounded to the nearest hundred thousand.

   (c) 350,000 is halfway between __300,000__ and 400,000. 350,000 is __400,000__ when rounded to the nearest hundred thousand.

   (d) 475,000 is __500,000__ when rounded to the nearest hundred thousand.

   (e) When rounding to the nearest hundred thousand, if the digit in the __ten thousands__ place is __5__ or more we round up to the next hundred thousand. When it is __4__ or less we round down.

Practice

2. Indicate the approximate location of 950,749 on each number line with an arrow.

   (a)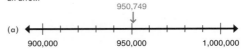

   950,749 is __1,000,000__ when rounded to the nearest hundred thousand.

   (b)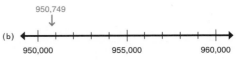

   950,749 is __950,000__ when rounded to the nearest ten thousand.

   (c)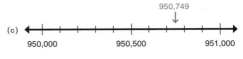

   950,749 is __951,000__ when rounded to the nearest thousand.

   (d)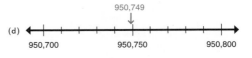

   950,749 is __950,700__ when rounded to the nearest hundred.

3. Round each number to the nearest hundred thousand.

   (a) 109,920 __100,000__
   (b) 639,501 __600,000__
   (c) 250,000 __300,000__
   (d) 97,522 __100,000__

4. The diameter of Jupiter at the equator is 142,984 km. Round this number to:

   (a) the nearest hundred thousand. __100,000__
   (b) the nearest ten thousand. __140,000__
   (c) the nearest thousand. __143,000__
   (d) the nearest hundred. __143,000__

Challenge

5. A number, when rounded to the nearest thousand, ten thousand, or hundred thousand is 500,000. What is the least whole number it could be?
   499,500

6. A number, when rounded first to the nearest thousand, then the nearest ten thousand, then the nearest hundred thousand, is 500,000. What is the least whole number it could be?
   444,500

Exercise 7 • pages 19–21

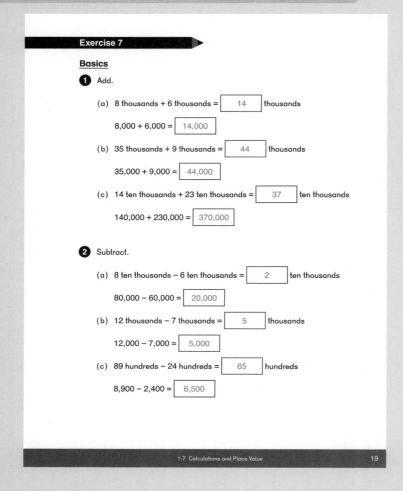

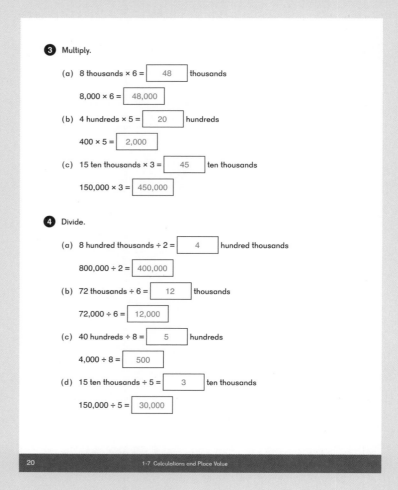

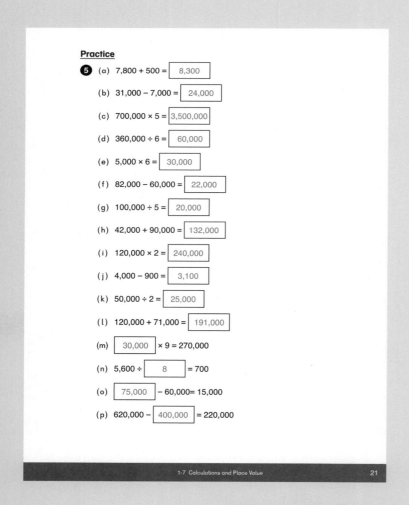

Exercise 8 • pages 22–24

Exercise 8

Check

1 (a) In 714,564, the digit 5 stands for 5 ___hundreds___.

(b) In the number for four hundred twenty-three thousand, sixty-seven, the value of the digit in the ten thousands place is ___20,000___.

2 (a) Write the numbers in numerals.

90,000 + 5,000 + 50 + 2	95,052
94 ten thousands + 80 hundreds + 70 tens	948,700
9 ten thousands + 7 thousands + 4 hundreds	97,400
The greatest 6-digit even number with 4 in the ten thousands place and all digits different.	948,762
The least 6-digit odd number with 9 in the hundred thousands place and 4 in the thousands place.	904,001

(b) Write the numbers above in order from greatest to least.

948,762   948,700   904,001   97,400   95,052

3 Continue the number patterns.

(a) **79,325** **79,225** 79,125 79,025 78,925

(b) **97,167** **98,167** 99,167 100,167 101,167

(c) **248,642** **248,742** 248,842 248,942 249,042

4 The diameter of Neptune at the equator is 49,532 km. Round this number to:

(a) the nearest ten thousand.   50,000

(b) the nearest thousand.   50,000

(c) the nearest hundred.   49,500

5 The diameter of Saturn at the equator is 120,536 km. Round this number to:

(a) the nearest hundred thousand.   100,000

(b) the nearest ten thousand.   120,000

(c) the nearest thousand.   121,000

6 Write >, <, or = in each ◯.

(a) 89,000 − 100 ⓘ> 9,000 × 9

(b) 3,400 + 800 ⓘ< 5 × 4,000

(c) 48,000 ÷ 8 ⓘ< 24,000 + 5,000

(d) 230,000 − 80,000 ⓘ> 62,000 + 37,000 + 8,000 + 13,000

7 A house costs 6 times as much as a car. The car costs $40,000. How much does the house cost?
40,000 × 6 = 240,000
$240,000

8 A number is 32,000 less than the quotient of 560,000 ÷ 7. What is the number?
560,000 ÷ 7 = 80,000
80,000 − 32,000 = 48,000
48,000

9 A number is rounded to 400,000 when rounded to the nearest ten thousand. What is the least possible and what is the greatest possible whole number it could be?
Least possible: 395,000
Greatest possible: 404,999

Challenge

10 There are 5,000 nails in five boxes. The first and second boxes have 2,700 nails altogether. The second and third boxes have 2,000 nails altogether. The third and fourth boxes have 1,800 nails altogether. The fourth and fifth boxes have 1,700 nails altogether. How many nails are in each box?

1st: 5,000 − 2,000 − 1,700 = 1,300
2nd: 2,700 − 1,300 = 1,400
3rd: 2,000 − 1,400 = 600
4th: 1,800 − 600 = 1,200
5th: 1,700 − 1,200 = 500

Check: 1,300 + 1,400 + 600 + 1,200 + 500 = 5,000

11 Four blocks, red, blue, green, and yellow, are in a row. The red block is to the right of the blue block, but not directly beside the blue block. The green block is not next to the blue block. The yellow block is next to the red block. What is the order of the blocks from left to right?
blue, yellow, red, green

Teacher's Guide 4A Chapter 1

Notes

Chapter 2 Addition and Subtraction — Overview

Suggested number of class periods: 6–7

	Lesson	Page	Resources	Objectives
	Chapter Opener	p. 45	TB: p. 37	Investigate addition and subtraction of five-digit numbers.
1	Addition	p. 46	TB: p. 38 WB: p. 25	Add five-digit numbers using a standard algorithm.
2	Subtraction	p. 49	TB: p. 42 WB: p. 28	Subtract five-digit numbers using a standard algorithm.
3	Other Ways to Add and Subtract — Part 1	p. 52	TB: p. 46 WB: p. 31	Use mental math to add and subtract up to five-digit numbers.
4	Other Ways to Add and Subtract — Part 2	p. 54	TB: p. 49 WB: p. 34	Use mental math to add or subtract a number close to a multiple of 10, 100, or 1,000.
5	Word Problems	p. 57	TB: p. 52 WB: p. 38	Solve multi-step word problems with addition and subtraction.
6	Practice	p. 59	TB: p. 55 WB: p. 42	Practice concepts from the chapter.
	Workbook Solutions	p. 61		

Chapter 2 Addition and Subtraction

Notes

In Dimensions Math 3A, students learned how to add and subtract four-digit numbers using the standard algorithm.

In this chapter, students will:

- Extend their knowledge of the addition and subtraction algorithm to five-digit numbers.
- Extend mental math strategies.
- Use rounding and strategic choice of a number to estimate values and find sums and differences.
- Apply these skills in multi-step word problems.

Addition and Subtraction Algorithms

The algorithm is important because of its simplicity. It requires only calculations involving addition and subtraction of one-digit numbers.

In Dimensions Math 3A, students used place-value discs to understand how to add and subtract with standard algorithms. Place-value discs are modeled in this chapter for visual reference. It is assumed that students already understand regrouping and have mastered the basic algorithms.

Students who need additional practice with addition and subtraction algorithms should do so with place-value discs and three or four-digit numbers.

For a more basic review of the addition and subtraction algorithms, reference Dimensions Math 3A Chapter 2 Lessons 1–3.

Other Methods of Addition and Subtraction

Mental math is performed using mental strategies that leverage number sense. In Dimensions Math 4A, students build on previous knowledge to make computation easier. Solving an expression using mental math does not mean that no part of the solution can be written, or that the computations can only be done orally.

Mental math is simply doing some part of the calculation mentally. The mental math strategies allow students to calculate more quickly.

Students should eventually be able to choose if they want to calculate mentally or not; they can use the algorithm if they are not certain their answer is correct or cannot think of a strategy immediately.

In Lessons 3 and 4, mental math strategies will be reinforced and extended to five-digit numbers.

For example:

- Subtracting from 10,000 using 10,000 = 9,000 + 900 + 90 + 10:

> Find the difference between 5,672 and 10,000.
>
> ```
>   5 thousands  6 hundreds  7 tens   2 ones
> + ? thousands  ? hundreds  ? tens   ? ones
>   9 thousands  9 hundreds  9 tens  10 ones
> ```

- Subtracting from 1,000 or a multiple of 100:

> ```
>    7,536 − 998              623 − 298
>     /    \                   /    \
>  6,536   1,000            323     300
>
>  1,000 − 998 = 2          300 − 298 = 2
>  6,536 + 2 = 6,538        323 + 2 = 325
> ```

- Over-adding and over-subtracting:

> ```
>         + 1,000         − 2
>  7,536 ────────▶ 8,536 ────▶ 8,534
>
>         − 1,000         + 2
>  7,536 ────────▶ 6,536 ────▶ 6,538
> ```

Encourage students to use mental math strategies for problems in future chapters, and continue to develop these strategies when opportunities arise.

Chapter 2 Addition and Subtraction

Notes

Bar Models

In Dimensions Math 3A, students learned how to draw a bar model to interpret and solve a two-step word problem. In this chapter, they interpret and apply their knowledge of model drawing to more complex word problems.

At this stage, it is expected that students have experience using bar models to help solve word problems. If students are unfamiliar with bar models, reference the following Dimensions Math 3A lessons.

- Bar models with addition and subtraction: Chapter 2 Lessons 8–11
- Bar models for multiplication and division: Chapter 4 Lessons 7–9

For quick reference, below is a review of the two main types of models students learned in Dimensions Math 3: Part-whole Models and Comparison Models.

Part-whole Models

This type of bar model extends the understanding of part-whole relationships from a number bond to a bar model. A bar model pictorially represents the relationship between the size of the quantities more proportionally.

There are 22,000 people at the concert.
There are 9,500 adults at the concert.
How many children are at the concert?

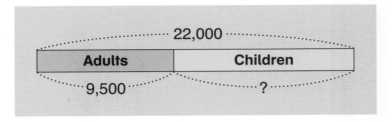

22,000 − 9,500 = ▢

or

9,500 + ▢ = 22,000

There are 12,500 children at the concert.

Comparison Models

This type of bar model allows students to compare quantities. They are particularly useful in representing multi-step problems.

There are 10,000 boys and 16,000 girls at the soccer match.

(a) How many fewer boys than girls are at the soccer match?
(b) How many children were at the soccer match in all?

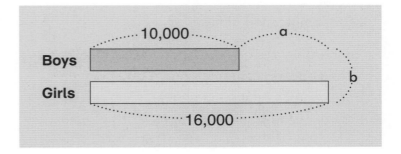

Comparison models reinforce the concepts of "more than" and "less than" and are related to the concept of difference.

Students can easily see that they need to subtract to find the answer to (a).

Although (b) is actually a part-whole type of question, the same model can be used to indicate the whole by placing a question mark at the end of the two bars.

The notation of a variable can be introduced simply by labeling the units in a bar model. In the previous example, the difference between boys and girls can be labeled "a," and the total number of boys and girls can be labeled "b."

To find the difference, students can write the equation: $a = 16{,}000 - 10{,}000$ or $10{,}000 + a = 16{,}000$.

To find the total number of children, students can write the equation: $b = 16{,}000 + 10{,}000$.

Chapter 2 Addition and Subtraction

Notes

In *Dimensions Math Textbook 4A*, students are not asked to use algebra to solve the problem, but teachers can show and explain how the bar model represents an equation.

Multi-step Models

In this chapter, students draw models for multi-step problems. As they solve harder problems, students will begin to appreciate the usefulness of the model drawing strategy.

For example, this problem from Lesson 5 **Do** may appear complicated at first:

An orchard harvested 3,300 pounds of Honeycrisp and Gala apples altogether, and 4,600 pounds of Gala and Fuji apples altogether. It harvested 1,900 pounds of Honeycrisp apples. How many more pounds of Fuji apples than Gala apples were harvested?

Once a bar model is drawn, students can easily see the steps to solve the problem.

- Subtract the weight of Honeycrisp apples from the combined weight of Honeycrisp and Gala apples to find the weight of Gala apples: $3,300 - 1,900 = 1,400$.

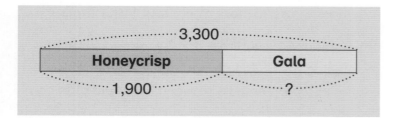

- Subtract the weight of Gala apples from the combined weight of Gala and Fuji apples to find the weight of Fuji apples: $4,600 - 1,400 = 3,200$.

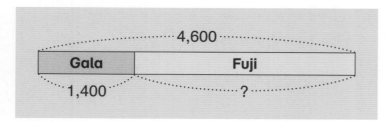

- Subtract the weight of Gala apples from the weight of Fuji apples to find how many more pounds of Fuji than Gala apples were harvested: $3,200 - 1,400 = 1,800$.

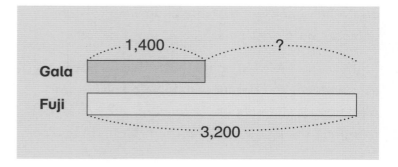

Models given in the textbook will be combined:

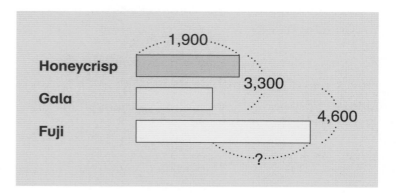

Students may find it useful to draw more than one model to solve the problem.

The purpose of drawing the models is not to encourage students to follow specific rules, but to understand the concepts and choose a good problem solving method. As students solve harder problems they will begin to appreciate the usefulness of the model drawing strategy.

When drawing bar models, students will also be expected to:

- Write an equation for each problem.
- Solve each problem. The answer should be given in a complete sentence. For example: "There were 1,800 more pounds of Fuji apples than Gala apples." This is to encourage students to not merely solve the problem correctly, but to solve the correct problem.

Chapter 2 Addition and Subtraction

Bar models are used for word problems throughout the Dimensions Math series. They are a tool to help students solve problems. While drawing models, students should be asked such questions as:

- "Which bar is longer? How do you know?"
- "What is known in the problem? What must we find?"
- "How do we know? What is unknown?"

Students can choose when to use bar models after completing the lessons that cover bar models. They should not need to draw them for simple one or two-step word problems, but should be encouraged to draw them for complex or multi-step problems.

Bar modeling is a good introduction to algebra as it provides pictorial representations of algebraic equations.

The notion of a variable can be introduced by labeling the units in the bar model. In the example on the previous page, the number of pounds of Gala apples can be labeled as "g." Students can write the equation:

$3,300 - 1,900 = g$ or $1,900 + g = 3,300$.
$g = 1,400$

The number of pounds of Fuji apples can be labeled as "f." Students can write the equation:

$4,600 - 1,400 = f$ or $1,400 + f = 4,600$.
$f = 3,200$

Finally, they will find the difference between f and g:
$3,200 - 1,400 = 1,800$.

Note: The numbers in bar model lessons are designed to reinforce mental math computation from earlier lessons. The focus of Chapter 2 is on the problem solving methods.

Notes & Materials

Materials

- 10-sided die
- Counters
- Dry erase markers
- Index cards
- Place-value discs
- Scratch paper
- Whiteboards

Blackline Masters

- Duel Scoring Sheet
- Differences within 100,000
- Number Cards
- Sums to 100,000
- Triangle Nim

Activities

Games and activities included in this chapter provide practice with addition and subtraction. They can be used after students complete the **Do** questions, or any time review and practice are needed.

Notes

Chapter Opener

Objective

- Investigate addition and subtraction of five-digit numbers.

Have students discuss the **Chapter Opener**. Using skills from Chapter 1, students may round to different places. For example, students might say:

- "I rounded 12,260 to 12,000, and 11,850 to 12,000, then added to get 24,000 gallons used."
- "I rounded both numbers to 10,000 and think they used more than 20,000 gallons."

★ Extend the activity by discussing how students could find about how much more water was used last month than this month. Students might share:

- "I rounded 12,260 to 12,300, and 11,850 to 11,800, then subtracted to find the difference of 500 gallons more."
- "I rounded both numbers to 12,000 and think they used the same amount of water both months. But that did not make sense, so I thought of 12,300 − 12,000 and think they used more than 300 gallons more last month."

This lesson may continue straight to Lesson 1.

Lesson 1 Addition

Objective
- Add five-digit numbers using a standard algorithm.

Lesson Materials
- Place-value discs

Think

Have students solve the **Think** problems and write an equation for each question. Students can use place-value discs, the algorithm, or any other known strategy to add the numbers together.

Ask students about the bar model shown in **Think**:

- "What information is given?"
- "How is the given information shown on the model?"
- "What is the question mark asking us to find?"

Have students who are unfamiliar with bar models think about the problem using smaller numbers. For example, if the car wash used 100 gallons of water in March and 20 more gallons of water in April than in March, how could we find the actual amount used in April?

Students can find the sum of the two numbers:

100 + 20 = 120

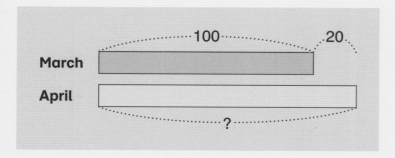

Then have students substitute the correct amounts from the problem:

95,617 + 25,186 = 120,803

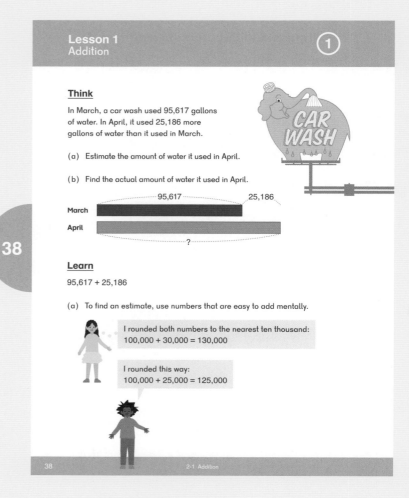

Record estimation and addition strategies on the board and discuss the methods students used.

Learn

Have students discuss the solutions that Sofia and Dion present and compare their own strategies with the ones in the textbook.

Sofia and Dion rounded to different place values.

Sofia rounded the numbers to the nearest ten thousand. Dion used a strategic choice of a number to add. He realizes 100,000 + 25,000 is easy to calculate mentally.

Discuss whose estimate is closer and why it is closer.

Another common student estimate might be:
95,000 + 25,000 = 120,000.

Mei introduces the sign ≈ for approximately equal to.

Point out to students that using an estimate can help them get a sense of the correctness of the computed answer.

Students should be fluent with the addition algorithm for four-digit numbers. Discuss how the algorithm is similar for five-digit numbers. There is an additional place value unit, hundred thousands. Point out that we regroup 10 ten thousands to 1 hundred thousand.

(b) Students can see from the picture that some of the place-value discs had to be regrouped into different places, which corresponds to the regrouping in the problem solved using the algorithm.

Do

1. Discuss the different estimates. Ask students whose estimate is closer and why.

   Alex asks students to evaluate whether their answers are reasonable. Students can compare their actual sums to their estimates.

2. Examples of possible student estimates:
   - 74,000 + 85,000 = 159,000
   - 75,000 + 85,000 = 160,000
   - 70,000 + 80,000 = 150,000
   - 100,000 + 100,000 = 200,000
   - 73,000 + 85,000 = 158,000

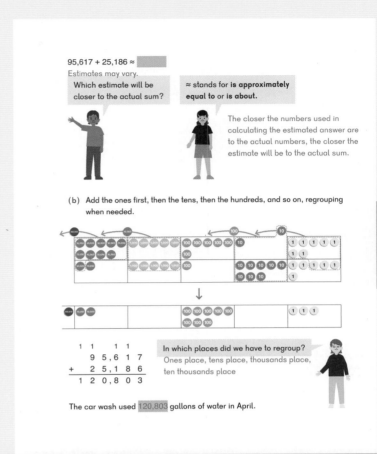

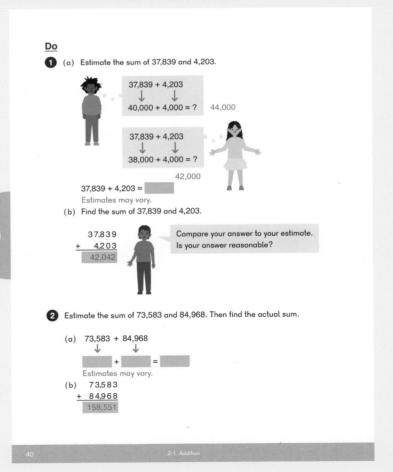

③ Discuss the different methods students used to estimate these large numbers. Emma rounds each number to the nearest 5,000 as she finds it easy to add the rounded numbers together.

Other possible student estimates:
- 76,000 + 27,000 + 5,000 = 108,000
- 80,000 + 30,000 + 5,000 = 115,000

④ Discuss some of the ways students rounded the numbers to estimate.

Activity

▲ **Sums to 100,000**

Materials: 10-sided die or Number Cards (BLM) 0–9, Sums to 100,000 (BLM)

Players take turns rolling the die or drawing a number card. On each turn, they place the number drawn in one of the empty boxes. Players add both five-digit numbers together.

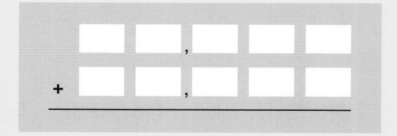

The player with the greatest final sum is the winner.

Exercise 1 • page 25

---

③ Estimate the sum of 75,583, 26,986, and 4,987. Then find the actual sum.
Estimates may vary.
(a) 75,583 + 26,986 + 4,987 ≈ ▇

I used these numbers: 75,000, 25,000, and 5,000, because they are easy to add.

(b) 75,583
    26,986
+   4,987
  107,556

④ Estimate the sums. Then find the actual sums.
Estimates may vary. Actual sums are provided.
(a) 62,387 + 698
63,085
(b) 8,421 + 43,875
52,296
(c) 75,284 + 6,396
81,680
(d) 82,087 + 29,903
111,990
(e) 62,805 + 86,102
148,907
(f) 148,135 + 52,009
200,144
(g) 459,242 + 240,989
700,231
(h) 3,895 + 987 + 9,213
14,095

⑤ 15,285 people went to the ballet in September, 12,378 people went in October, and 13,698 people went in November.
(a) About how many people went to the ballet in all three months?
Estimates may vary.
(b) Exactly how many people went to the ballet in all three months?
41,361 people

Exercise 1 • page 25

2-1 Addition   41

Lesson 2 Subtraction

Objective
- Subtract five-digit numbers using a standard algorithm.

Lesson Materials
- Place-value discs

Think

Have students solve the **Think** problems and write an equation for each question. Students can use place-value discs, the algorithm, or any other known strategy to add the numbers together.

Record their estimation and subtraction strategies on the board and discuss the methods students used.

Learn

Have students discuss the solutions that Alex and Mei present and have students compare their own strategies with the ones in the textbook.

Alex and Mei rounded to different place values. Discuss whose estimate is closer to the actual difference.

Alex rounded the numbers to the nearest ten thousand. Mei rounds 71,543 to the nearest thousand (72,000). She rounds 58,637 to the nearest ten thousand (60,000) to make a simple calculation. Mei knows 72 − 60 = 12, so 72,000 − 60,000 = 12,000.

Students should be fluent with the subtraction algorithm for four-digit numbers. Discuss how the algorithm is similar for five-digit numbers.

(b) Students can see from the picture that some of the place-value discs had to be regrouped into different places, which corresponds to the regrouping in the problem solved using the algorithm.

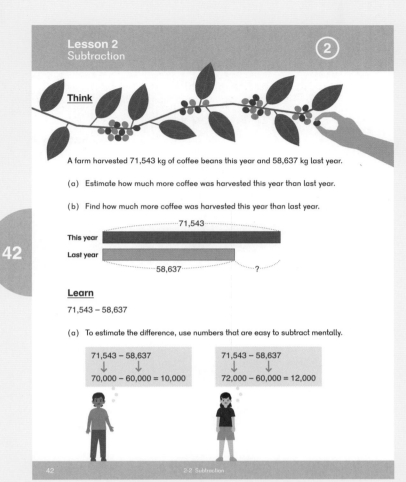

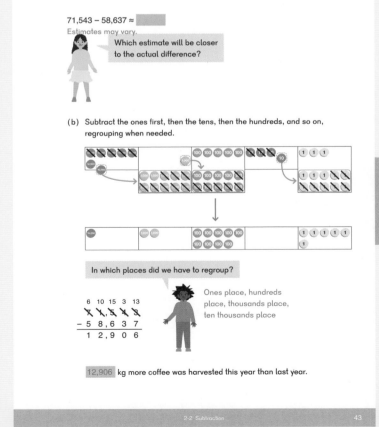

Do

① Discuss the different estimates. Ask students whose estimate is closer to the actual difference and why.

Mei asks students to evaluate whether or not their answers are reasonable. Students can compare the actual difference to their estimates.

② Possible student estimates:

- 80,000 − 27,000 = 53,000
- 80,000 − 30,000 = 50,000
- 80,000 − 25,000 = 55,000

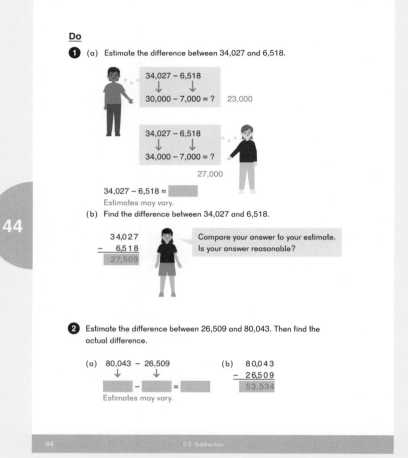

❸ Discuss the different methods students used to estimate.

Possible student estimates:

- 65,000 − 1,000 = 64,000
- 65,000 − 500 = 64,500
- 65,000 − 800 = 64,200
- 70,000 − 800 = 69,200

❹ Have students share how they rounded the numbers to estimate for some of the problems.

❺ (b) Discuss the bar model with students. Students may add the amounts spent on office furniture and computer equipment first, then subtract that amount from 50,000. They may also subtract one item from the total first and then subtract the other from that answer:

50,000 − 15,925 = 34,075
34,075 − 5,289 = 28,786
$28,786

Activity

▲ **Least**

Materials: 10-sided die or Number Cards (BLM) 0−9, Differences within 100,000 (BLM)

Players take turns rolling the die or drawing a number card. On each turn, they place the number drawn in one of the empty boxes.

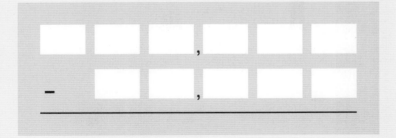

The player with the least difference is the winner.

Exercise 2 • page 28

Lesson 3 Other Ways to Add and Subtract — Part 1

Objective

- Use mental math to add and subtract up to five-digit numbers.

Lesson Materials

- Number Cards (BLM)

Think

Provide students with a set of cards from the **Think** activity. Have them find the pairs that add up to 10,000 and discuss patterns that they find.

Ask students, "What did you notice about the sum of the places in each pair of numbers that add to 10,000?"

Learn

Discuss the methods presented by the friends. Have students compare their own methods with the methods shown in the textbook.

Do

1. Students will apply Emma's strategy to subtraction. How many more ones can be added to 2 ones to get a total of 10 ones? (8)

   Ask students how the mental calculation is similar to the algorithm.

   When Mei regroups across the zeros in the thousands, hundreds, tens, and ones places, each place has a value of 9, with the exception of the ones, which has a value of 10.

   Mei knows 9,990 + 10 = 10,000. This allows her to subtract mentally from left to right.

2. Have students look for patterns in the problems. If students struggle ask, "How does solving problem (a) help when solving problems (b) and (c)?" The same question can be asked about each of the rows.

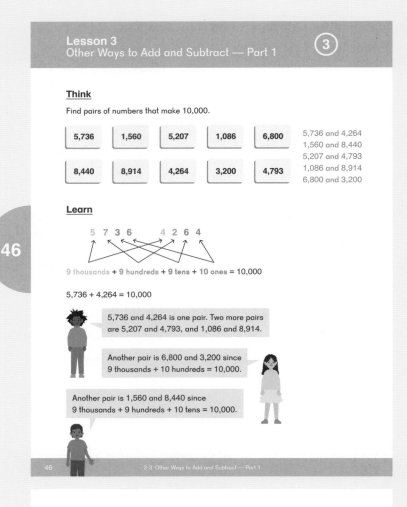

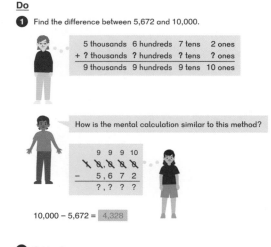

52   Teacher's Guide 4A Chapter 2   © 2019 Singapore Math Inc.

3) Emma decomposes, or splits, 70,000 into 60,000 and 10,000. She uses mental math to subtract 4,530 from 10,000, then adds that difference to 60,000.

4) Dion uses a similar strategy as Emma. He splits 8,500 into 8,400 and 100 and subtracts 62 from 100, then adds that difference to 8,400.

5) Have students share some of the methods they used to solve the problems.

Activity

▲ Match or Memory

Materials: Index cards

Provide students with a set of index cards and have them write 5 pairs of numbers that each sum to 10,000, similar to the **Think** activity.

Students can exchange or combine card decks with each other and play Match or Memory.

★ To extend, have students make cards with three numbers that add to 10,000.

Exercise 3 • page 31

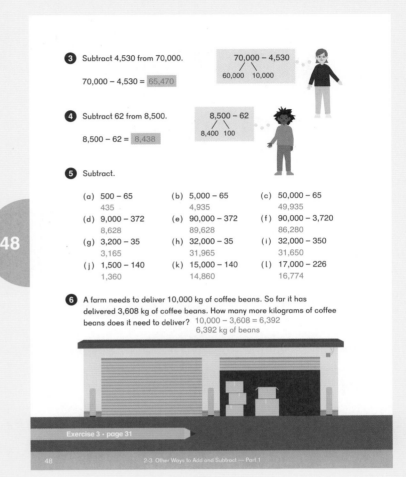

© 2019 Singapore Math Inc.   Teacher's Guide 4A Chapter 2   53

Lesson 4 Other Ways to Add and Subtract — Part 2

Objective

- Use mental math to add or subtract a number close to a multiple of 10, 100, or 1,000.

Think

Show students the problem:

464 + 298 = ?

Ask:

- "What is special about a number like 298 that makes it a good candidate for a mental math strategy?"
- "Can you recall methods we have already learned to add the numbers using mental math?"

Pose the **Think** problem and have students share how they solved it. Record their strategies on the board and discuss the methods used.

Learn

Discuss the methods shown in **Learn** and have students compare their own methods with the methods shown in the textbook.

Method 1

Alex splits 1,745 into 1,743 and 2.
998 + 2 = 1,000

Method 2

Sofia solves the problem by over-adding. She adds 1,000, which is 2 too many, then subtracts 2.

Note that both Methods 1 and 2 involve adding or subtracting 2. Alex splits 2 from the first number. Sofia adds 1,000 first, then subtracts 2.

Method 3

The standard algorithm can also be used.

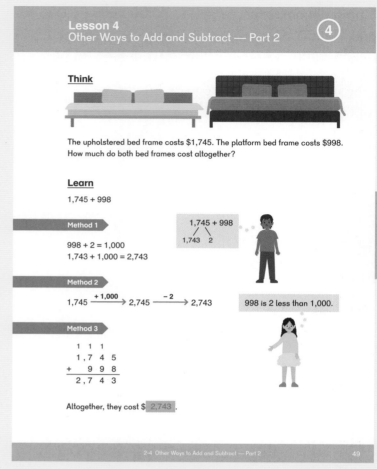

Do

① Alex, Mei, and Dion show three different ways to solve this problem.

Alex uses the strategy from Lessons 2–3.

Mei uses the standard algorithm.

Dion over-subtracts, similar to Method 2 in **Think**. He subtracts 1,000, which is 2 too many and adds the 2 back.

② Have students look for patterns in the problems. If students struggle ask, "How does solving problem (a) help when solving problems (b) and (c)?" The same can be asked about each row, as each row builds on the first problem of the row.

Have students share the methods they used to solve some of the problems. Encourage students to calculate using mental math.

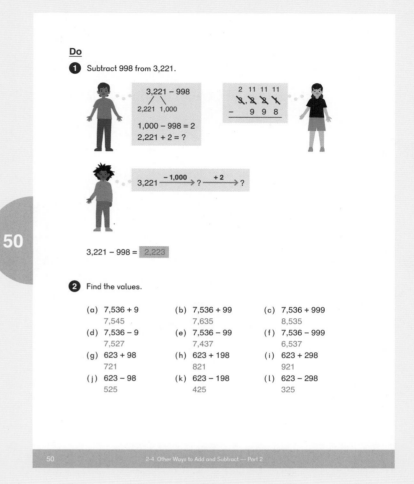

③ Emma splits 846 into 844 and 2 to perform a mental calculation. She takes 2 from 846 to add 58 to make 60, then solves 844 + 60.

Students might also split 58 into 54 and 4 and add 846 + 54 + 4.

④ Alex splits 703 into 663 and 40 as he knows he can easily subtract 39 from 40.

In both ③ and ④ the algorithm is also an efficient strategy.

⑤ Have students look for patterns in the problems and share the methods they used to solve some of the problems. If students need additional help ask, "How does solving problem (a) help when solving problems (b) and (c)?"

Activity

▲ Duel

Materials: Scratch paper or whiteboards, Duel Scoring Sheet (BLM)

Each player comes up with 3 numbers that add up to 1,000. When both players have chosen their numbers, they fill in the number duel boxes.

The player with the greater number gets a point.

For example, Player One chooses 800, 100, 100. Player Two chooses 500, 400, 100.

Player One gets 1 point: 800 > 500. Player Two gets 1 point: 400 > 100. No points are awarded for a tie.

After five rounds, the player with the most points wins.

★ To extend, students receive 1 bonus point for numbers chosen that are not multiples of one hundred.

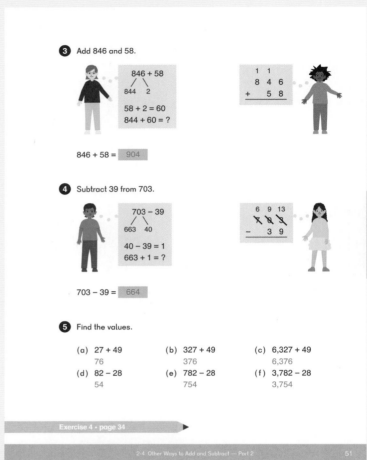

Exercise 4 • page 34

Lesson 5 Word Problems

Objective

- Solve multi-step word problems with addition and subtraction.

Think

Pose the **Think** problem and have students draw a bar model. When students need additional help with drawing bar models, prompt them with questions:

- "What quantities are given?"
- "How do we know which bar to draw longer?"
- "How many bars do we need to draw?"
- "Do we know how long to draw the bar for children? Does it matter in finding a solution?"
- "What can we find first?"

Have students use their bar models to solve the problem.

Learn

Discuss Emma's comment about her bar model.

From the model, students can see that they need to add 4,200 to 12,870 to find the number of men first. Then they can find the number of men and women and subtract that amount from the total number of people to find how many children are at the game.

Students may draw two sets of models. One finds the number of men and a second model finds the number of children.

Solving for the number of men:

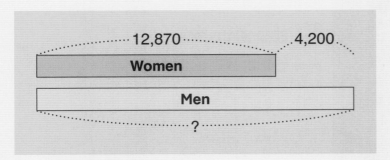

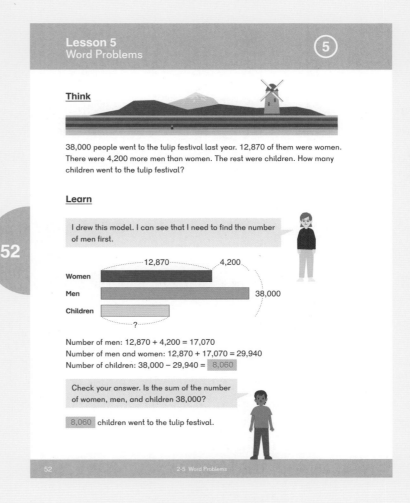

Solving for the number of children:

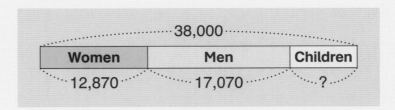

Have students share the methods they used to add and subtract.

A common error on this problem is for students to draw a part-whole bar model with women, men, and children as the three parts. Students may then end up with an incorrect answer if they think of the number of men as 4,200.

Students can also solve the problem using equations, with letters standing for the unknown quantities:

m = number of men   m = 12,870 + 4,200
c = number of children   c = 38,000 − 12,870 − m

© 2019 Singapore Math Inc.   Teacher's Guide 4A Chapter 2   57

Do

The bar models are given for the **Do** problems. Students should be encouraged to use mental math to solve the problems. If students need practice drawing bar models, present the problems on a whiteboard without the textbook pictures of the models.

❷ – ❸ If students are struggling with the absence of numbers on the bar models in the textbook, have them redraw the models with bars labeled with the related amounts.

❹ Students should see that the total number of cattle does not change.

Students can see from the model that to find the original number of cattle in the meadow at first (Before model), they need to look at the After model. They can subtract the number of cattle that moved from the field to the meadow (560) from the number of cattle in the meadow (1,690) to find how many cattle were in the meadow at first: 1,690 − 560 = 1,130.

Students can use the Before model and see that they should add 1,130 to the difference between the number of cattle in the field and in the meadow (1,240) to find the number of cattle in the field at first:

1,130 + 1,240 = 2,370

Exercise 5 • page 38

Lesson 6 Practice

Objective

- Practice concepts from the chapter.

After students complete the **Practice** in the textbook, have them continue to practice using mental math by playing games from this chapter.

5
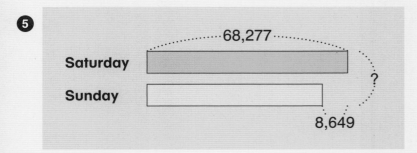

6
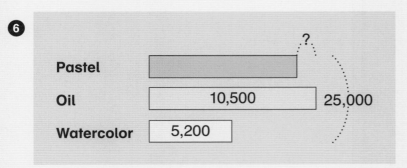

7
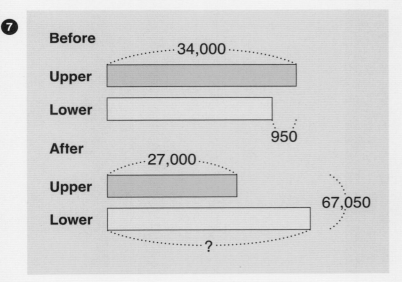

Exercise 6 • page 42

Brain Works

★ Nim

Materials: Counters, Triangle Nim (BLM)

Classic Nim: Lay out 15 counters in rows of 3, 5, and 7.

Two players take turns removing 1, 2, or 3 counters from any one row.

The player who forces the other player to take the last counter is the winner.

Try changing the number of counters in each row. Does it affect the outcome?

Questions to ask:

- "Is there a way to always win?"
- "Does it matter who goes first?"
- "What happens if the number of counters is changed?"

★ Triangle Nim

Materials: Counters, Triangle Nim (BLM)

Place one counter on each triangle on the game board.

Two players take turns removing 1 or 2 counters from two adjacent triangles.

The player who takes the last counter is the winner.

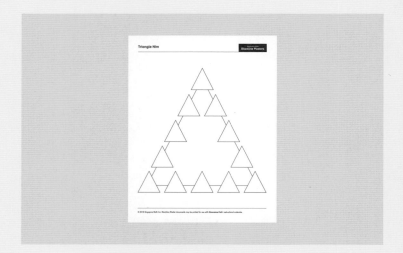

Exercise 1 • pages 25–27

Chapter 2 Addition and Subtraction

Exercise 1

Basics

1 (a) Estimate the sum of 85,459 and 8,586 two different ways as shown.

$$85,459 + 8,586 \qquad\qquad 85,459 + 8,586$$
$$\downarrow \qquad\qquad\qquad\qquad \downarrow$$
$$90,000 + 10,000 = \boxed{100,000} \qquad 85,000 + 9,000 = \boxed{94,000}$$

(b) Which estimate will be closer to the actual sum?
94,000

(c) Which estimate was easier to calculate mentally?
100,000

(d) Find the sum of 85,459 and 8,586.

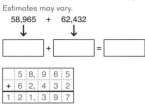

2 Estimate and then find the sum of 58,965 and 62,432.
Estimates may vary.

58,965 + 62,432

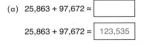

Practice

3 (a) Estimate the sum of 462,754 and 63,689 two different ways as shown.

$$462,754 + 63,689 \qquad\qquad 462,754 + 63,689$$
$$\downarrow \qquad\qquad\qquad\qquad \downarrow$$
$$500,000 + 60,000 = \boxed{560,000} \qquad 460,000 + 60,000 = \boxed{520,000}$$

(b) Find the sum of 462,754 and 63,689.

	4	6	2,	7	5	4
+		6	3,	6	8	9
	5	2	6,	4	4	3

4 Estimate and then find the sum of 34,842 + 5,783 + 7,874.
Estimates may vary.

34,842 + 5,783 + 7,874

□ + □ + □ = □

	3	4,	8	4	2
		5,	7	8	3
+		7,	8	7	4
	4	8,	4	9	9

5 How many digits are in the sum of 73,921 and 82,692?
Estimate: Adding the 7 and 8 in the ten thousands place gives a 2-digit sum, so there will be one more digit in the sum.

6 digits

6 Circle the number that is equal to 5,968 + 58,934 + 4,982 without calculating the exact answer.

75,904 14,984 **69,884** 158,644

7 Estimate and then find the sum. Estimates may vary.

(a) 25,863 + 97,672 ≈ □

25,863 + 97,672 = 123,535

(b) 186,485 + 41,264 + 786 ≈ □

186,485 + 41,264 + 786 = 228,535

Challenge

8 In the following problem, the letters D, E, and Y stand for different digits.
What is the number DYE?

```
  1 1 1
  E D D Y
+   Y Y Y
  D E E D
```

DYE = 685   even, not 0

There is regrouping from the hundreds place to the thousands place, so E + 1 = D. The sum in the hundreds and tens place is the same, so there is regrouping to the hundreds place as well: D + Y + 1 = 1E. Similarly, Y + Y = 1D. D must be even, and cannot be 0, so must be 2, 4, 6 or 8. Use trial and error for Y = 6, 7, 8 and 9. Only 8 works. Students may also simply start with trial and error where D is even and not 0.

Exercise 2 • pages 28–30

Exercise 2

Basics

1 (a) Estimate the difference between 62,342 and 8,724 two different ways as shown.

62,342 − 8,724
↓   ↓
60,000 − 9,000 = 51,000

62,342 − 8,724
↓   ↓
62,000 − 9,000 = 53,000

(b) Which estimate will be closer to the actual difference?
53,000

(c) Which estimate was easier to calculate mentally?
51,000

(d) Find the difference between 62,342 and 8,724.

```
  6 2,3 4 2
−   8,7 2 4
  5 3,6 1 8
```

2 Estimate and then find the difference between 51,582 and 38,958.
Estimates may vary.

51,582   −   38,958
  ↓   ↓

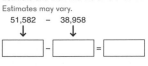

```
  5 1,5 8 2
− 3 8,9 5 8
  1 2,6 2 4
```

Practice

3 (a) Estimate the difference between 852,065 and 57,393 two different ways as shown.

852,065 − 57,393
↓   ↓
900,000 − 60,000 = 840,000

852,065 − 57,393
↓   ↓
850,000 − 50,000 = 800,000

(b) Find the difference between 852,065 and 57,393.

```
  8 5 2,0 6 5
−   5 7,3 9 3
  7 9 4,6 7 2
```

4 Estimate and then find the difference between 35,974 and 75,506.
Estimates may vary.

75,506   −   35,974
  ↓   ↓

[   ] − [   ] = [   ]

```
  7 5,5 0 6
− 3 5,9 7 4
  3 9,5 3 2
```

5 Is 624,987 − 78,965 closer to 540,000 or 500,000?
540,000

6 How many digits are in the value of 150,747 − 83,965?
5 digits

7 Circle the number that is equal to 214,094 − 78,934 − 42,762 without calculating the exact answer.

| 80,218 | 92,398 | 92,001 | 106,498 |

8 Estimate and then find the difference.
Estimates may vary.

(a) 80,703 − 72,389 ≈ [   ]

80,703 − 72,389 = 8,314

(b) 416,415 − 41,267 ≈ [   ]

416,415 − 41,267 = 375,148

Challenge

9 In the following problem, the letters A, B, C, D, and E stand for different digits. Find two possible values for ABCDE.

```
  A B A A B
−     C B C
  D C E D 1
```

Since AB changes to DC after subtraction, B must be 0 and C must be 9. Try 2, 3, etc. for A. Two possible solutions are:

```
  2 0 2 2 0        3 0 3 3 0
−     9 0 9     −     9 0 9
  1 9 3 1 1        2 9 4 2 1
ABCDE = 20,913    ABCDE = 30,924
```

Exercise 3 • pages 31–33

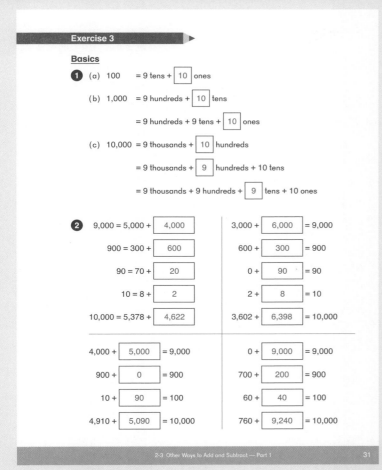

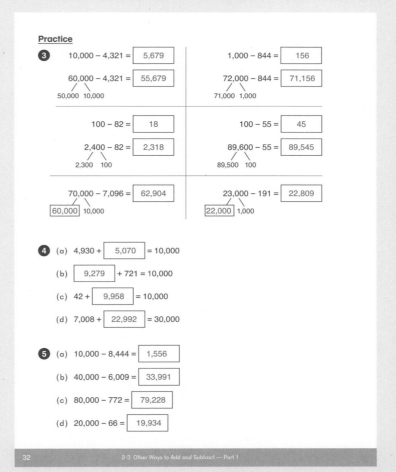

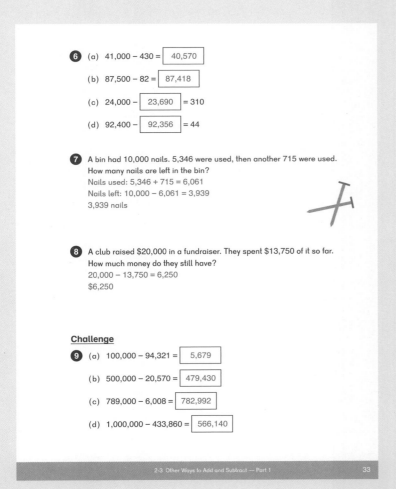

Exercise 4 • pages 34–37

Exercise 4

Basics

1 5,784 + 998 = 5,782 + [1,000] = [6,782]
   5,782 2

   5,784 —−2→ [5,782] —+1,000→ [6,782]
   5,784 —+1,000→ [6,784] —−2→ [6,782]

2 5,784 − 998 = 4,784 + [2] = [4,786]
   4,784 1,000

   5,784 —+2→ [5,786] —−1,000→ [4,786]
   5,784 —−1,000→ [4,784] —+2→ [4,786]

3
   4,327 + 99 = [4,426]   4,327 − 99 = [4,228]
   4,326 1   4,227 100

   4,327 + 299 = [4,626]   4,327 − 299 = [4,028]
   4,326 1   4,027 300

   4,327 + 1,999 = [6,326]   4,327 − 1,999 = [2,328]
   4,326 1   2,327 2,000

4 (a) 50 + 60 = [110]
   (b) 450 + 60 = [510]
   (c) 7,450 + 60 = [7,510]
   (d) 7,458 —+60→ [7,518] —−1→ [7,517]
   (e) 7,458 + 59 = [7,457] + 60 = [7,517]
       7,457 1

5 (a) 21 − 4 = [17]
   (b) 210 − 40 = [170]
   (c) 3,210 − 40 = [3,170]
   (d) 3,218 —−40→ [3,178] —+2→ [3,180]
   (e) 3,218 − 38 = [3,178] + 2 = [3,180]
       3,178 40

6 (a) 4,387 —+40→ [4,427] —−1→ [4,426]
       4,387 + 39 = [4,426]
   (b) 6,228 —−40→ [6,188] —+1→ [6,189]
       6,228 − 39 = [6,189]

Practice

7 Add or subtract.

8,432 + 99 8,531 **C**	9,007 − 98 8,909 **H**	10,000 − 8,492 1,508 **T**
3,889 + 998 4,887 **M**	9,868 − 999 8,869 **V**	6,129 + 498 6,627 **R**
984 + 799 1,783 **S**	4,725 + 89 4,814 **G**	10,000 − 398 9,602 **N**
9,000 − 998 8,002 **I**	10,000 − 28 9,972 **K**	7,407 + 1,998 9,405 **O**
9,921 − 2,999 6,922 **E**	987 + 78 1,065 **P**	10,000 − 3,702 6,298 **A**

Write the letters that match the answers and learn a fun fact.

S	H	A	R	K	S		C	A	N
1,783	8,909	6,298	6,627	9,972	1,783		1,873	8,531	6,298	9,602

N	E	V	E	R		S	T	O	P
9,602	6,922	8,869	6,922	6,627		8,431	1,783	1,508	9,405	1,065

		M	O	V	I	N	G
8,969	1,873	4,887	9,405	8,869	8,002	9,602	4,814	8,431	8,969

Challenge

8
5,400 + 800 = [6,200]	4,200 − 600 = [3,600]
57 + 7 = [64]	81 − 5 = [76]
5,457 + 807 = [6,264]	4,281 − 605 = [3,676]
25,000 + 8,000 = [33,000]	46,000 − 9,000 = [37,000]
213 + 8 = [221]	775 − 8 = [767]
25,213 + 8,008 = [33,221]	46,775 − 9,008 = [37,767]

9 (a) 3,864 + 606 = [4,470]
   (b) 9,472 − 508 = [8,964]
   (c) 72,509 + 6,005 = [78,514]
   (d) 13,381 − 4,007 = [9,374]

10 399 + 498 + 597 + 696 = [2,190]

   400 + 500 + 600 + 700
   − 1 − 2 − 3 − 4

11 30,022 + 29,998 + 29,980 + 30,010 = [120,010]

   30,000 + 30,000 + 30,000 + 30,000
   + 22 − 2 − 20 + 10

Teacher's Guide 4A Chapter 2 © 2019 Singapore Math Inc.

Exercise 5 • pages 38–41

Exercise 5

Basics

1 For each problem, finish labeling the bar model with the information given in the problems. Mark the quantity that needs to be found with a question mark. Then solve the problem.

(a) There are 3 laptops for sale at a store. Laptop A costs $1,200 and is $900 less than Laptop B. Laptop C costs $285 more than Laptop B. How much do all three laptops cost altogether?

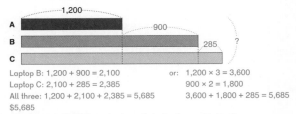

Laptop B: 1,200 + 900 = 2,100   or: 1,200 × 3 = 3,600
Laptop C: 2,100 + 285 = 2,385   900 × 2 = 1,800
All three: 1,200 + 2,100 + 2,385 = 5,685   3,600 + 1,800 + 285 = 5,685
$5,685

(b) A bookstore sold 435 fewer books on Saturday than on Friday. On Sunday, it sold 3,260 books, which was 1,420 more books than on Friday. How many books did it sell altogether during the three days?

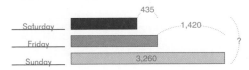

Friday: 3,260 − 1,420 = 1,840
Saturday: 1,840 − 435 = 1,405
All three: 3,260 + 1,840 + 1,405 = 6,505
6,505 books

(c) There were 8,435 more kilograms of beans in Bin A than in Bin B in a warehouse. After 2,630 kilograms of beans were taken out of Bin A, 11,550 kilograms of beans were left in Bin A. How many kilograms of beans were in Bin B?

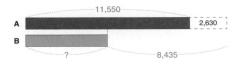

Bin A (before): 11,550 + 2,630 = 14,180
Bin B: 14,180 − 8,435 = 5,745
5,745 kg

(d) There were 456 fewer students in School A than in School B. After 162 students transferred from School B to School A, how many more students were in School B than School A?

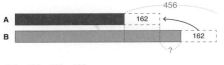

456 − 162 − 162 = 132
132 more students

Practice

2 4,693 concert tickets were sold on Tuesday. 1,840 more concert tickets were sold on Monday than on Wednesday. 1,524 fewer concert tickets were sold on Tuesday than on Wednesday. How many concert tickets were sold on Monday?

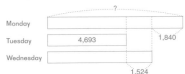

Wednesday: 4,693 + 1,524 = 6,217
Monday: 6,217 + 1,840 = 8,057
8,057 concert tickets

3 String A is 560 cm shorter than String C and 120 cm longer than String B. String C is 2,320 cm long. How long are Strings A and B together?

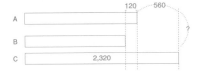

String A: 2,320 − 560 = 1,760
String B: 1,760 − 120 = 1,640
1,760 + 1,640 = 3,400
3,400 cm

4 There were 1,430 trees in an orchard. 750 of them were apple trees and the rest were peach trees. After 82 of the apples trees died, and 35 more peach trees were planted, how many more peach trees than apple trees were there?

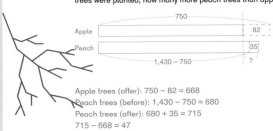

Apple trees (after): 750 − 82 = 668
Peach trees (before): 1,430 − 750 = 680
Peach trees (after): 680 + 35 = 715
715 − 668 = 47
47 more peach trees

5 There were 1,350 cars in Lot A. After 420 cars were moved from Lot B to Lot A, there were 1,720 cars in Lot B. How many cars are in both lots altogether?

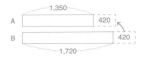

Lot A: 1,350 + 420 = 1,770
Both lots: 1,770 + 1,720 = 3,490
3,490 cars

Exercise 6 • pages 42–46

Exercise 6

Check

1 Estimate and then find the values.   Estimates may vary.

(a) 45,846 + 32,185 ≈ ☐

45,846 + 32,185 = 78,031

(b) 345,000 − 83,723 ≈ ☐

345,000 − 83,723 = 261,277

(c) 62,089 + 490,076 ≈ ☐

62,089 + 490,076 = 552,165

(d) 87,002 − 49,725 ≈ ☐

87,002 − 49,725 = 37,277

2 (a) 4,892 + 99 = 4,991   (b) 10,000 − 985 = 9,015

(c) 70,000 − 4,568 = 65,432   (d) 892 + 498 = 1,390

(e) 7,718 + 89 = 7,807   (f) 6,500 − 87 = 6,413

(g) 82,780 − 998 = 81,782   (h) 84,345 + 142 = 84,487

3 There were 5,878 passengers on a ship. At the end of a week-long cruise, 4,341 people got off, and the rest stayed on for the next week. 4,362 more people got on for the next week.

(a) Estimate how many people are now on the ship.
Estimates will vary.

(b) Find how many people are now on the ship.
5,878 − 4,341 = 1,537
1,537 + 4,362 = 5,899
5,899 people

(c) The capacity of the ship is 6,000 passengers. By how much is the ship under capacity for the second week?
6,000 − 5,899 = 101
101 people

4 On a cruise ship, 479,314 gallons of fresh water were consumed in one day. 13,174 of those gallons were frozen for ice. How many gallons of water were not used for ice?
479,314 − 13,174 = 466,140
466,140 gallons

5 A cruise ship needs to have at least 40,000 pounds of potatoes for a two-week cruise. It has 2,786 pounds left over from the last trip. It can order potatoes from a supplier in amounts rounded to the nearest thousand pounds. How many pounds of potatoes must it order?
40,000 − 2,786 = 37,214
38,000 pounds

6 For a one-week cruise, a cruise ship had 3,450 more pounds of white flour than whole wheat flour and 1,250 pounds less rye flour than whole wheat flour. It had 2,240 pounds of rye flour. How many pounds of the three types of flour did it have altogether?

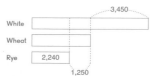

Wheat: 2,240 + 1,250 = 3,490
White: 3,490 + 3,450 = 6,940
All three: 6,940 + 3,490 + 2,240 = 12,670
12,670 pounds

7 On a cruise, there were 3,935 adults. There were 1,980 fewer children than adults. After 2,560 adults and 1,420 children got off the ship for the day at one stop, how many passengers were left on the ship?

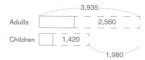

Adults (after): 3,935 − 2,560 = 1,375
Children (before): 3,935 − 1,980 = 1,955
Children (after): 1,955 − 1,420 = 535
Passengers (after): 1,375 + 535 = 1,910
1,910 passengers

Challenge

8 There are 1,000 flags in a row. From left to right, the 60th to 290th flags are blue. From right to left, the 150th to 410th flags are blue. The rest of the flags are red. How many red flags are there?
60th to 290th flags: 290 − 60 + 1 = 231
150th to 410th flags: 410 − 150 + 1 = 261
Blue flags: 231 + 261 = 492
Red flags: 1,000 − 492 = 508
508 red flags

9 15 digits are in a row. The sum of every 3 consecutive digits is 10. What are the values of P and Q?

Start with the 8. The two numbers to the left of 8 could be 1 and 1 or 0 and 2. The pattern repeats, since every 3 consecutive digits must have a sum of 10. Q must be 8 as well as the digit after P. If the first digit is 1, the two digits to the left of 8 must be 1 and 1, and P must therefore be 1.
P = 1, Q = 8

10 In the following problem, the letters X, Y, and Z stand for different digits. What is the greatest possible sum?

```
      1
  X 2 Y Z        1 2 2 4
  Z Y 3 X        4 2 3 1
  Y Z X 4        2 4 1 4
+ 2 X Y Z      + 2 1 2 4
  -------        -------
  9 9 9 3        9 9 9 3
```

X = 1
Y = 2
Z = 4

The greatest sum would have as many digits 9 as possible, starting from the left. In the thousands place, if X + Y + Z + 2 = 9, then X + Y + Z = 7. The same would be true for the hundreds place. The only possibilities for X, Y, and Z are 1, 2, and 4, and there is no regrouping from the tens place. Use trial and error for X, Y, or Z being 1, 2, or 4 in the ones and tens place. The greatest possible sum is 9,993.

11

		Example
Step 1	Pick any 4-digit number where the thousands digit and hundreds digit are different.	4,632
Step 2	Reverse its digits.	2,364
Step 3	Find the difference between these two numbers.	4,632 − 2,364 = 2,268
Step 4	Reverse its digits.	8,622
Step 5	Find the sum of the difference and its reverse.	2,268 + 8,622 = 10,890

Do these 5 steps with other numbers. What do you observe?
The final number is always either 10,890 or 9,999.

12

		Example
Step 1	Use any three different digits to make the greatest and least 3-digit number.	743 and 347
Step 2	Find the difference between them.	743 − 347 = 396
Step 3	Continually repeat steps 1 and 2 using the digits from this new number.	
		963 − 369 = 594
	What happens?	954 − 459 = 495
		954 − 459 = 495

You eventually get 495 again and again.
Try some other numbers. Does the same thing happen?
Yes.

Notes

Chapter 3 Multiples and Factors — Overview

Suggested number of class periods: 6–7

	Lesson	Page	Resources	Objectives
	Chapter Opener	p. 73	TB: p. 57	Investigate factors and multiples.
1	Multiples	p. 74	TB: p. 58 WB: p. 47	Find multiples of a given one-digit number.
2	Common Multiples	p. 77	TB: p. 61 WB: p. 50	Find common multiples.
3	Factors	p. 81	TB: p. 65 WB: p. 53	Find factors of whole numbers up to 100. Determine if a one-digit number is a factor of another number.
4	Prime Numbers and Composite Numbers	p. 84	TB: p. 68 WB: p. 57	Identify prime and composite numbers.
5	Common Factors	p. 87	TB: p. 72 WB: p. 60	Identify common factors of two whole numbers.
6	Practice	p. 90	TB: p. 76 WB: p. 63	Practice topics from the chapter.
	Workbook Solutions	p. 92		

Chapter 3 Multiples and Factors

In Dimensions Math 3A, students learned multiplication facts up to 10 × 10.

In this chapter, students will learn about:
- Multiples and common multiples
- Factors and common factors
- Prime and composite numbers

Factors and multiples help students develop their understanding of properties of whole numbers, and facilitate simplifying fractions and finding common denominators when adding fractions.

Students should be able to recall their multiplication and division facts up to 10 × 10 quickly and easily. Remediation should be provided for students who are weak on these facts. To review these skills or find games for fact practice, refer to Dimensions Math 3A Chapter 4 and Dimensions Math 3B Chapter 8.

Multiples

A multiple is a name given to the product of a given number and a whole number. In this book, we only consider whole number multiples of whole numbers. Multiples of 6, for example, are 6, 12, 18, 24, etc. There are an infinite number of multiples of any given number.

A number is always the first multiple of itself because any number times 1 is the number: 1 × 6 = 6, so 6 is the first multiple of 6.

18 is a multiple of 6 because three sixes make 18. Students will list some multiples and discern if a number is a multiple using division. A multiple of a number can be divided by that number with no remainders. Students will use this to determine if a number is a multiple of another number.

For example, 78 is a multiple of 6 because 78 ÷ 6 = 13. There is no remainder.

Some rules of divisibility will be discovered by marking multiples on a hundred chart. Students will discover rules of divisibility by marking multiples on a hundred chart. These rules make it easier to identify factors through simple elimination, rather than dividing each number to check if there are remainders.

Rules of divisibility include:

Multiples of 2 have 0, 2, 4, 6, and 8 in the ones place.

Every even number is a multiple of 2, therefore every even number is divisible by 2.

Multiples of 5 have a 0 or 5 in the ones place.

Every number with a 0 or 5 in the ones place is divisible by 5.

Multiples of 3 have digits that sum to 3.

The digits of 54 sum to 9. 9 is divisible by 3, so 54 is divisible by 3. This is helpful with larger numbers. For example, the digits in 594 sum to 18. The digits 1 and 8 sum to 9, and 9 is divisible by 3, so 594 is divisible by 3.

Multiples of 9 have digits that sum to 9.

The digits of 81 sum to 9, so 81 is divisible by 9. The digits of 378 sum to 18, and the digits of 18 sum to 9, so 378 is divisible by 9. Note that while this rule works for multiples of 3 and 9, it does not work for multiples of other numbers. All multiples of 9 are multiples of 3, but not all multiples of 3 are multiples of 9. For example, 75 is a multiple of 3, but not 9.

Common Multiples

Students will find common multiples of numbers by listing and comparing the multiples of each number.

Multiples of 3: 3, 6, 9, **12**, 15, 18, 21, **24**, 27

Multiples of 4: 4, 8, **12**, 16, 20, **24**, 28

12 and 24 are the first two common multiples of 3 and 4.

An easy way to find a common multiple is to multiply the numbers together. For example, 12 is a common

Chapter 3 Multiples and Factors

multiple of 3 and 4: 3 × 4 = 12. Every multiple of 12 is a multiple of both 3 and 4, because 12 is a multiple of both 3 and 4.

Factors

Students will apply their knowledge of multiples to factors. A factor is the whole number, that when multiplied by a number, yields the multiple:

- 18 is a multiple of 3, because 6 × 3 is 18. 3 is a factor of 18, because 3 × 6 is 18.
- 18 is a multiple of 6 because 3 × 6 is 18. 6 is a factor of 18 because 6 × 3 is 18.

There is a finite number of factors of a given number. A number is always a factor of itself, because itself × 1 is itself.

To find all of the factors of 18, we can first list them as pairs, beginning with 1:

1 × 18
2 × 9
3 × 6

Students should list the pairs systematically. In doing so, they can use divisibility rules to see that they do not need to test if 4 and 5 are factors. 6 is already listed in 3 × 6, so students know that they have found all of the factor pairs.

Students will then list factors in order from least to greatest: 1, 2, 3, 6, 9, 18.

Common Factors

Students will find common factors of two or more numbers:

Factors of 18: **1, 2, 3, 6,** 9, 18
Factors of 12: **1, 2, 3,** 4, **6,** 12

1, 2, 3, and 6 are common factors of 12 and 18.

Common factors can also be found using division and divisibility rules. Since both 12 and 18 are even numbers and even numbers are multiples of 2, we know that 2 is a common factor of 12 and 18.

Both 12 and 18 are multiples of 3, so 3 is a common factor of 12 and 18.

1 is always a common factor of any two numbers, but 0 is not a factor of any number as 0 times anything results in 0. For example, 0 × 18 = 0, so 0 is not a factor of 18.

Students often confuse the terms "multiples" and "factors." Have students practice using the terms often.

Note: While students will look for the first common multiple, at this level students will not be asked to find least common multiples or greatest common factors.

Prime and Composite Numbers

If a number has only two factors, 1 and itself, the number is called a prime number. 7 is a prime number as its only factors are 1 and 7.

Numbers with more than two factors are called composite numbers. The number 8 is a composite number as its factors are 1, 2, 4, and 8.

The numbers 0 and 1 are neither prime nor composite since a prime number has exactly two factors. The number 1 has only one factor and therefore does not meet the definition of a prime number or composite number. The number 0 has an infinite number of factors because any number multiplied by 0 is 0.

Chapter 3 Multiples and Factors

Materials

- 10-sided die, 2
- Colored pencils
- Counters
- Deck of cards with face cards removed
- Dry erase markers
- Dry erase sleeves
- Flash cards with a multiplication expression on one side and the product on the other side
- Markers
- Square tiles, 12 for each student
- Two-color counters
- Whiteboards

Blackline Masters

- Graph Paper
- Hundred Chart
- Number Cards
- Numbers to 40 Chart — 1 Start
- Shut the Box — Factors

Activities

Games and activities included in this chapter provide practice with factors and multiples. They can be used after students complete the **Do** questions, or any time review and practice are needed.

Chapter Opener

Objective

- Investigate factors and multiples.

Have students discuss the **Chapter Opener**.

The purpose of this **Chapter Opener** is to review missing factor problems that students learned in Dimensions Math 3A. They should be reminded that they can divide to find the missing factor.

Students that need to review math facts can make flash cards or practice with the activities below.

Activities

- **Salute!**

Materials: Deck of cards with face cards removed

Salute! is played with three students. The Caller shuffles, then deals out the deck to two players. The third player is the Caller.

When the Caller says, "Salute!" the other players place the top cards from their piles on their foreheads. The two players can see each other's cards, but not their own.

The Caller tells the players the product of the two numbers on their cards.

The player who says the number on his card first is the winner.

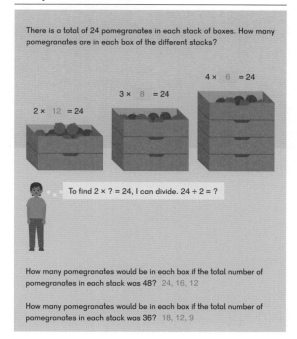

- **Facts Forwards and Backwards**

Materials: Flash cards with a multiplication expression on one side and the product on the other side

Partners take turns drawing a fact card. One student looks at the side of the card with the expression and the other player looks at the side of the card with the product.

The player looking at the expression asks, "Is the product 54?"

The partner says "yes" or "no" and tries to guess the expression. "Is the equation 7 × 8?" Player One says, "No." Player Two asks, "Is it 6 × 9?" Player One replies, "Yes."

© 2019 Singapore Math Inc.   Teacher's Guide 4A Chapter 3   73

Lesson 1 Multiples

Objective

- Find multiples of a given one-digit number.

Lesson Materials

- Dry erase sleeves
- Hundred Chart (BLM)

Think

Discuss the **Think** task and ask students to consider the term "multiples." Ask them what they think the term means based on its use here.

Have students find answers to the **Think** questions and share how they solved them. Record their strategies on the board and discuss the methods used.

Learn

Ensure students understand that any whole number (other than 0) multiplied by 4 is a multiple of 4. A multiple of a given number is a product of that given number and any whole number (other than 0).

Have students compare their own strategies with the ones in the textbook.

Mei points out that division can be used to find if 76 is a multiple of 4. 76 is a multiple of 4 if 76 divides evenly by 4.

Students can use mental calculation by splitting 76 into 40 and 36:

$$76 \div 4 = 10 + 9$$
$$\swarrow \qquad \searrow$$
$$40 \qquad 36$$

$76 \div 4 = (40 + 36) \div 4$
$\qquad = (40 \div 4) + (36 \div 4)$
$\qquad = 10 + 9$
$\qquad = 19$

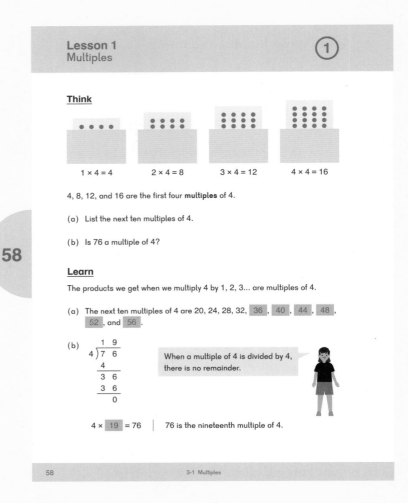

When listing all of the multiples of 4, 76 is the nineteenth multiple because 19 fours makes 76, or $19 \times 4 = 76$.

Do

❸ Prompt students with these questions:

(a)
- "Is there a number that can be multiplied by 7 to get 91?"
- "How many sevens make 91?"

Students can use the division algorithm or a mental calculation:

(b)
- "Is there a number that can be multiplied by 8 to get 91?"
- "How many eights make 91?"

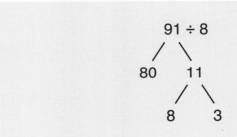

❹ Students may see that 64 is an even number. All even numbers are multiples of 2. This problem can be extended by asking, "Is 64 a multiple of 32? Is it a multiple of 16?"

❺ Similar to the **Learn** example, students can find the fourth multiple of 9 by multiplying 4 × 9.

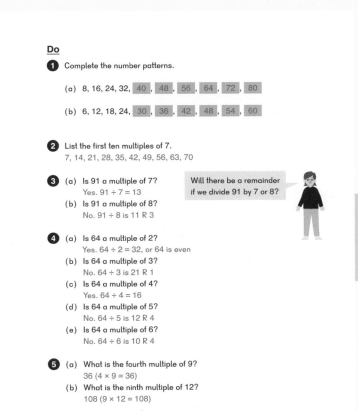

6 Students are discovering some basic rules of divisibility.

(a) Students should mark all multiples of 2 and see that they are all even numbers (they all end in 0, 2, 4, 6, or 8). They can erase the chart before marking multiples of 5 for (b).

(b) Students should see that multiples of 5 have either a 0 or 5 in the ones place.

(c), (d) Have students mark both multiples of 3 and 9 on the Hundred Chart (BLM) together using two different colors. Ask students, "What do you notice about the multiples of 9?" (They are also multiples of 3.)

Students should realize that adding the digits together does not work for other numbers. The digits of 25 add to 7, however, 7 is not divisible by 5. Likewise, the digits of 23 add to 5, but 23 is not divisible by 5.

Activity

▲ Five in a Row — Multiples

Materials: Hundred Chart (BLM), dry erase sleeve, Number Cards (BLM) 1–10

Players take turns drawing a card and marking any open multiple of the number on the card. Because all numbers are multiples of the number 1, the 1 card is essentially a wild card.

The winner is the first player to mark five numbers in a row horizontally, vertically of diagonally.

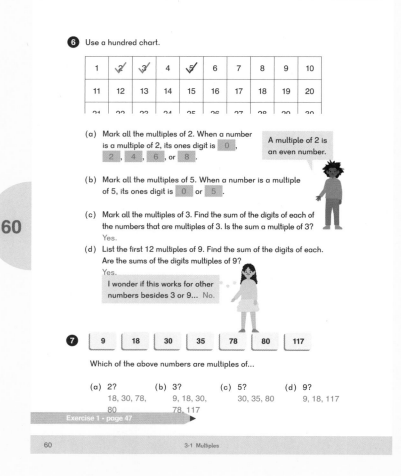

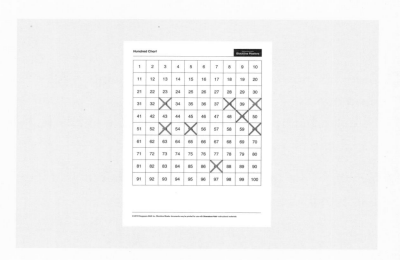

Exercise 1 • page 47

Lesson 2 Common Multiples

Objective
- Find common multiples.

Lesson Materials
- Colored pencils
- Hundred Chart (BLM)
- Two-color counters

Think

Pose the **Think** problem and have students work on it independently. Provide students with two-color counters and colored pencils and have them show or draw their solutions.

Ensure that students understand the goal is to buy enough pots and seedlings so that each seedling has a pot and there are no leftover seedlings or pots.

Students may draw a solution similar to the **Think** image by thinking, "If I have 1 box of pots and 1 tray of seedlings…"

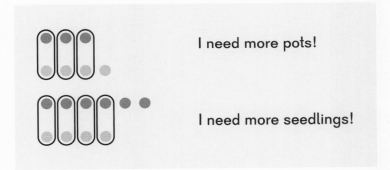

I need more pots!

I need more seedlings!

Continuing until they see:

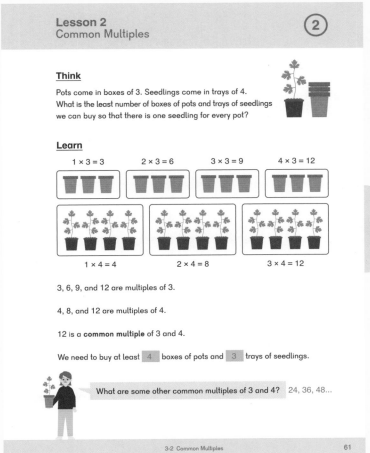

Learn

Have students compare their solutions from **Think** with the ones shown in the textbook.

Explain that we can find common multiples by listing multiples of numbers until we find one that both numbers have in common.

Multiples of 3: 3, 6, 9, **12**
Multiples of 4: 4, 8, **12**

The first multiple of 3 that is also a multiple of 4 is 12.

When answering Emma's question, students may see that the common multiples of 3 and 4 are also multiples of 12, or 3 × 4.

To extend, ask students what other solutions are possible. They could buy 8 boxes of pots and 6 trays of seedlings to have a total of 24 of each, and so on.

Do

❶ (b) Have students write the given multiples and circle the common multiples.

❷ When students start by listing the multiples of the greater number, they do not need to list multiples of the smaller number if they recognize a multiple of the larger number is also a multiple of the smaller number.

Students should recall that all even numbers are multiples of 2, therefore, all even multiples of 7 are also multiples of 2.

❸ Even multiples of 7: 14, 28, 42, etc.

Using Mei's strategy, students should start by listing multiples of 10. Multiples of 10 are also multiples of 5. By finding multiples of 3 that have a 0 in the ones place, we can find common multiples of all three given numbers.

Do

❶ (a) Find the first two common multiples of 4 and 6.

　　Multiples of 4: 4, 8, ⑫, 16, 20, ㉔...

　　Multiples of 6: 6, ⑫, 18, ㉔, 30, 36...

　　12 and 24 are the first two common multiples of 4 and 6.

(b) What are some other common multiples of 4 and 6?
　36, 48, 60...

❷ Find the first four common multiples of 2 and 7.

Instead of listing multiples of both numbers, I could just list multiples of the greater number, and check if they are multiples of 2 as well.

Multiples of 7: 7, 14, 21, 28, 35, 42, 49, 56...

The first four common multiples of 2 and 7 are 14, 28, 42, and 56.

❸ Find a common multiple of 3, 5, and 10.   30

The multiples of 10 are 10, 20, 30, 40, 50...
30 is also a multiple of 3 and 5.

④ Have students mark the multiples as listed on a Hundred Chart (BLM). They can use different colored pencils for each multiple and each shape if that makes it easier for them to see common multiples.

Example:

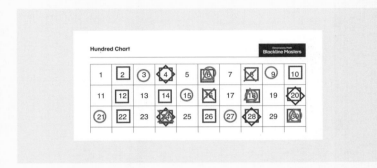

Activity

▲ What Number Am I? Riddles

- I am between 60 and 70. I am a multiple of 8. 64
- I am a multiple of both 2 and 3 between 20 and 30. 24
- I am a multiple of both 2 and 7. I am a two-digit number. The sum of my digits is 17. 98
- I am a three-digit number. 2, 3, and 41 are my factors. 246
- 3 and 9 are two of my factors. 108 and 162 are two common multiples of 27 and me. The digit in my tens place exceeds the digit in my ones place by 1. 54
- I am a common multiple of 3, 6, and 8. The sum of my digits is 9. I am a three-digit number. 144
- I am a common multiple of 4, 5, and 7. The digit in my tens place exceeds the digit in my hundreds place by 6, and the digit in my ones place by 8. I am a three-digit number. 280
- 84 and I are common multiples of 2, 3, and 7. The digit in my ones place is $\frac{1}{2}$ the digit in my tens place. 42
- 40, 80, and I are common multiples of 5 and 8. The sum of my digits is 3. I am a three-digit number. 120

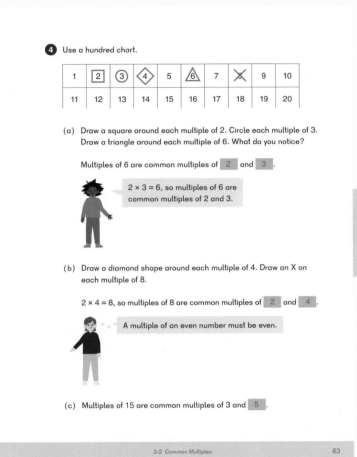

5 Students can use the multiplication tables of 6 and 9 for all numbers except 108.

Here, students could:

- Divide 108 by 6 and by 9 to see if 108 is a common multiple of 6 and 9.
- Realize that 54 is a common multiple of 6 and 9. Since 108 is a multiple of 54, it must also be a common multiple of 6 and 9.

6 Students may see that the common multiples of 3 and 7 are:

3 × 7, 6 × 7, 9 × 7 since 3, 6, 9 are all multiples of 3.

7 Students should see that multiples of 8 are always multiples of 4. They can then look for common multiples of 6 and 8.

Multiples of 6: 6, 12, 18, **24**, 30, 36, 42, **48**
Multiples of 8: 8, 16, **24**, 32, 40, **48**

Once students find 24, they will see that all common multiples of 6 and 8 are multiples of 24.

8 – 9 Students should relate these questions to the problem in **Think**. They can list multiples of each number to find common multiples.

Activity

▲ Five in a Row — Common Multiples

Materials: Hundred Chart (BLM), dry erase sleeves, Number Cards (BLM) 1–10 or 2 ten-sided dice

Students take turns rolling the dice or drawing two cards. On each turn, they put a mark on any open common multiple of their die roll. For example, if students roll an 8 and a 3, they may mark any common multiple: 24, 48, 72, 96.

The winner is the first player to mark five numbers in a row horizontally, vertically or diagonally.

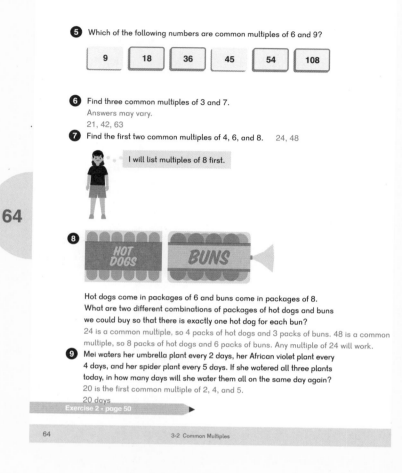

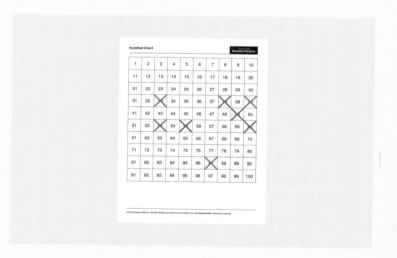

Exercise 2 • page 50

Lesson 3 Factors

Objectives

- Find factors of whole numbers up to 100.
- Determine if a one-digit number is a factor of another number.

Lesson Materials

- Graph Paper (BLM)
- Square tiles, 12 for each student

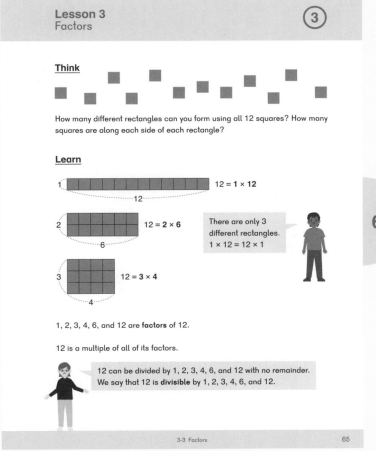

Think

Provide students with square tiles and Graph Paper (BLM). Have them record the different rectangles they can make with the tiles on the Graph Paper.

Discuss student strategies for solving the problem.

Learn

Have students compare their solutions from **Think** with the ones shown in the textbook. Introduce the term "factor": any whole number can be expressed as the product of two or more factors.

When students count how many tiles are on each side, they can multiply the sides together and get the total number of squares. The total is a multiple of both those side numbers.

Students should see that if all of the equations are listed, the factors are also listed. Listing the factors helps students keep their thoughts organized.

1 × 12
2 × 6
3 × 4

Note that factors are typically listed from least to greatest: 1, 2, 3, 4, 6, 12.

Help students relate multiples and factors:

- 12 is a multiple of each factor of 12. For example, 1 is a factor of 12, and 12 is a multiple of 1.
- 4 is a factor of 12, and 12 is a multiple of 4.

Discuss Emma's comment on divisibility.

Do

❷ Students may need to systematically list factors. In this problem, the factors 1, 2, and 3 are already given and students find their factor pairs.

Note that here and in ❺, factors are found as their pairs and then listed from least to greatest.

Discuss Emma's question. Students should see that all factors greater than 6 have already been found. They only need to check numbers between 3 and 6:

- We know that 4 × 4 is 16, and 18 is only 2 more than 16, so 4 is a not a factor of 18.
- The number 18 does not end in 0 or 5, so we know 5 is not a factor of 18.

This may be fairly easy to see with the number 18, however, it will become helpful when finding all the factors of greater numbers, like 60, which has 12 factors.

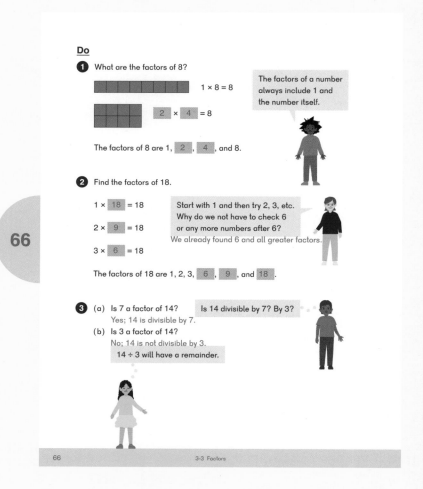

④ Students can check to see if the numbers are divisible by 6 (i.e. 16 ÷ 6 = 2 remainder 4), or they can list the multiples of 6: **6**, 12, 18, 24, 30, **36**, 42, 48, **54**, **60**.

Students should know that 6 is a factor of 6.

⑤ Mei reminds us that a multiple of an even number is always even. We do not need to check to see if 6, 8, 10, 12, or 14 are factors of 75. Numbers greater than 15 do not need to be checked since they have already been found.

If this is confusing, have students consider the products of one odd and one even number. For example, 3 × 2 is 6, 5 × 2 is 10, etc. In each case, even though one of the factors is odd, the product is even.

⑥ Students may need to list the factors as pairs first, then make a list of the factors from least to greatest.

Activity

▲ Factor Game

Materials: Numbers to 40 Chart — 1 Start (BLM) in dry-erase sleeve, markers

On each turn, players choose and cross off a number on the Numbers to 40 Chart — 1 Start (BLM). That number is the player's score for that round.

After Player One crosses off a number, Player Two then marks all factors of that number and adds them together to get her score for the round.

Example: Player One (red) chooses 21 and records that score. Player Two (blue) finds all of the remaining factors of 21 (1, 3, 7) and crosses them off the game board. She adds them to get her score for the round (11).

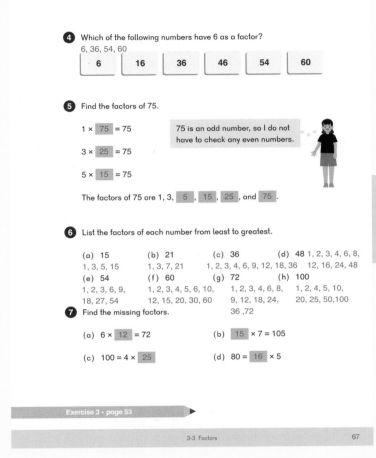

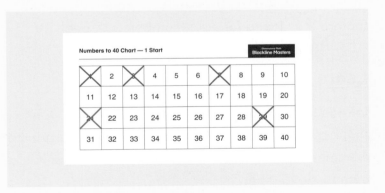

On the next round, Player Two chooses 29 and adds that to her score of 11.

Player One finds the factors of 29: 1, 29. Both 1 and 29 are already crossed off, so she scores 0 for this round.

Play ends when there are no more numbers to cross off.

Exercise 3 • page 53

Lesson 4 Prime numbers and Composite Numbers

Objective
- Identify prime and composite numbers.

Lesson Materials
- Dry erase sleeves
- Graph Paper (BLM)
- Hundred Chart (BLM)
- Square tiles, 12 for each student

Think

Provide students with square tiles and Graph Paper (BLM). Have them make rectangles with 2, 3, 4, and up to 12 tiles and record the different rectangles they can make with the tiles on the Graph Paper (BLM).

Discuss what students notice about the numbers:

- 2 tiles can only make a 1 × 2 rectangle.
- 3 tiles can only make a 1 × 3 rectangle.
- 4 tiles can make 1 × 4 and 2 × 2 rectangles.
- 5 tiles can only make a 1 × 5 rectangle.
- 6 tiles can make 1 × 6 and 2 × 3 rectangles.

Discuss **Think** and have students count the number of factors they found for each number, 2 through 12.

Learn

Introduce the terms "prime" and "composite." A number that has only 1 and itself as factors is a prime number. A number that has factors other than 1 and itself is a composite number.

Have students identify the prime numbers they found in **Think** within 12.

Have students answer Dion's and Emma's questions.

Alex and Sofia share two additional rules relating to prime and composite numbers. The number 1 is neither prime nor composite as it has only 1 factor. The number 0 has an infinite (or uncountable) number of factors.

★ Possible extension: Ask students why all even numbers greater than 2 are composite.

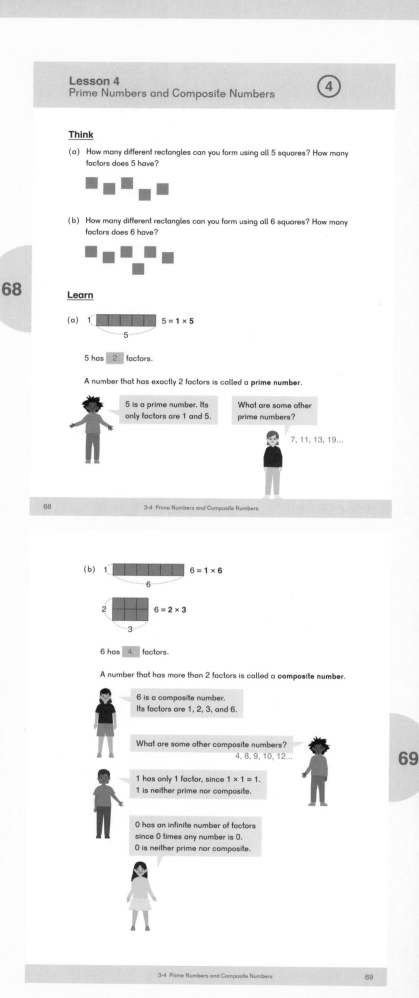

Do

2 (a) Students should recall that if the sum of the digits of a number is divisible by 3, then the number is divisible by 3. For example, with the number 39, 3 + 9 = 12, which is divisible by 3; therefore, 39 must have a factor pair of 3 and another number (13).

(b) Students can quickly see that 47 is not even and is not a multiple of 5.

The digits sum to 11, which is not divisible by 3 or 9. If a number is not divisible by 2 then it cannot be divisible by 4, 6, or 8.

Students may then divide 47 by 7 or 11 and see that 47 does not divide by any of the numbers without a remainder.

Do

1 Copy and complete the table.

Number	Factors	Prime	Composite
2	1, 2	✓	
3	1, 3	✓	
4	1, 2, 4		✓
7	1, 7	✓	
8	1, 2, 4, 8		✓
9	1, 3, 9		✓
10	1, 2, 5, 10		✓
11	1, 11	✓	
12	1, 2, 3, 4, 6, 12		✓

2 (a) Is 39 prime or composite?
Composite
(b) Is 47 prime or composite?
Prime

3 Which of the following numbers are prime numbers?

15   13   21   53   565   1,312

④ (a) Students should see that they do not need to cross off multiples of 4 or 6 as they are even and have already been crossed off. The numbers 2, 3, 5, and 7 should be circled.

To answer Mei's question, students should see that the next prime number is 11, and 11 × 11 is greater than 100. They do not have to check any other numbers as the rest are prime.

Have students circle the remaining prime numbers on Hundred Chart (BLM).

Activity

▲ Shut the Box — Factors

Materials: Shut the Box — Factors (BLM), several sets of Number Cards (BLM) 0—9, 10 counters

Players take turns drawing 2 cards.

The player forms a two-digit number with the cards she drew. She covers any factor of that two-digit number with a counter on her game board. If the number is already covered, her turn is over.

The first player to cover all of the numbers on her game board is the winner.

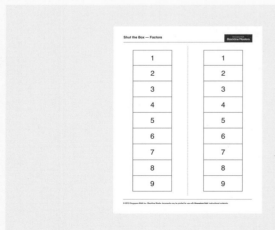

Exercise 4 • page 57

Lesson 5 Common Factors

Objective

- Identify common factors of two whole numbers.

Lesson Materials

- Two-color counters, 40 for each pair of students

Think

Provide students with two-color counters to represent the pink and blue erasers. Ensure students understand that they need to put only one color in each box, and each box needs to have the same amount of erasers. There can be as many boxes as they need.

Discuss student strategies for solving the **Think** problem.

Learn

Ensure students have found all the possible combinations. Have students sort the counters similarly to how they are sorted in the textbook. They should see that:

- If there is 1 eraser in each box, there is 1 pink eraser in each of 12 boxes and 1 blue eraser in each of 18 boxes.
- If there are 2 erasers in each box, there are 2 pink erasers in each of 6 boxes and 2 blue erasers in each of 9 boxes.
- If there are 3 erasers in each box, there are 3 pink erasers in each of 4 boxes and 3 blue erasers in each of 6 boxes.
- If there are 6 erasers in each box, there are 6 pink erasers in each of 2 boxes and 6 blue erasers in each of 3 boxes.

Students may have tested other numbers while solving the **Think** problem. They would find that they could not put 12 pink erasers and 18 blue erasers into boxes of 4 or 5 since 4 and 5 are not common factors of 12 and 18.

Lesson 5
Common Factors

Think

Sofia and Alex are putting 12 pink erasers and 18 blue erasers into boxes.
- Each box must have the same number of erasers.
- Each box must have only one color of erasers.
- All the erasers should be used.

How many erasers could they put in each box?

Learn

$1 \times 12 = 12$ | $1 \times 18 = 18$

They could put 1 eraser in each box.

$2 \times 6 = 12$ | $2 \times 9 = 18$

They could put 2 erasers in each box.

When students see the factors listed, it is easy to see that 1, 2, 3, and 6 are factors of both 12 and 18. They are "common factors."

Discuss Sofia's questions about 4 erasers in each box.

Do

① Alex finds factor pairs for each number.

② If students have trouble seeing the common factors, have them write the factors of each number from least to greatest.

Factors of 24: **1, 2, 3, 4, 6**, 8, **12**, 24
Factors of 36: **1, 2, 3, 4, 6**, 9, **12**, 18, 36

③ Mei suggests checking if both numbers are divisible by 7 to see if 7 is a common factor of 42 and 86.

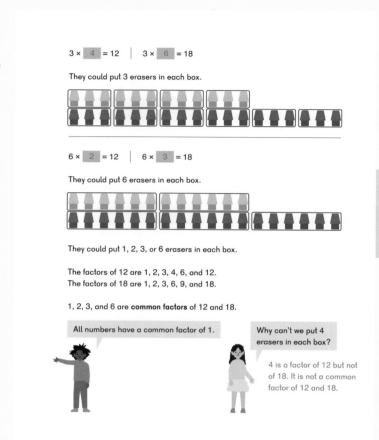

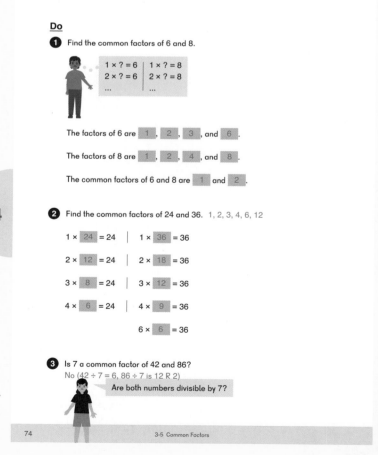

④ Students should check for common factors by dividing both numbers by the numbers given, 4 and 8.

⑤ (a) Multiples of 5 have a 0 or 5 in the ones place. Only even numbers are divisible by 2, so only numbers ending in 0 will have a common factor of 2 and 5.

(b) While students can list all factors of 3 and 9 to find common factors, they should recall that if the sum of the digits of a number is divisible by 3, then the number is divisible by 3. The same is true of 9. The sum of the digits in 54 (5 + 4) and 90 (9 + 0) are divisible by both 3 and 9. 54 and 90 have common factors of 3 and 9.

⑦ Students should relate this problem to the problem in **Think**.

- They can make 1 bouquet with 24 red roses and 32 yellow roses in each.
- They can make 2 bouquets with 12 red roses and 16 yellow roses in each.
- They can make 4 bouquets with 6 red roses and 8 yellow roses in each.
- They can make 8 bouquets with 3 red roses and 4 yellow roses in each.

The greatest factor that 24 and 32 have in common is 8.

Exercise 5 • page 60

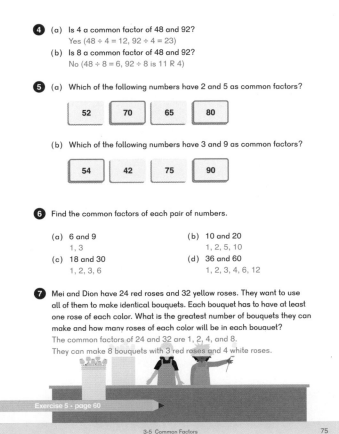

Lesson 6 Practice

Objective

- Practice topics from the chapter.

After students complete the **Practice** in the textbook, have them continue finding factors and multiples by playing games from this chapter.

Lesson 6 Practice P 6

1. (a) List the first ten multiples of 7.
   7, 14, 21, 28, 35, 42, 49, 56, 63, 70
   (b) List the first twenty multiples of 3.
   3, 6, 9, 12, 15, 18, 21, 24, 27, 30, 33, 36, 39, 42, 45, 48, 51, 54, 57, 60
   (c) Which of the numbers found in (a) and (b) are common multiples of 3 and 7?
   21, 42

2. What are the first four common multiples of 2, 3, and 4?
   12, and multiples of 12, are common multiples of 2, 3, and 4.
   12, 24, 36, 48

3. Find all the factors of each number.
   (a) 20
   1, 2, 4, 5, 10, 20
   (b) 28
   1, 2, 4, 7, 14, 28
   (c) 50
   1, 2, 5, 10, 25, 50
   (d) 64
   1, 2, 4, 8, 16, 32, 64

4. Which of the following numbers are multiples of 4?
   | 64 | 78 | 92 | 100 |

5. Which of the following numbers are multiples of 7?
   | 28 | 67 | 84 | 96 |

76   3-6 Practice

6. Which of the following numbers have 3 as a factor?
   | 18 | 27 | 83 | 91 | 99 | 118 |

7. Which of the following numbers are composite?
   | 18 | 27 | 83 | 91 | 99 | 118 |

8. List the nine prime numbers that are less than 25.
   2, 3, 5, 7, 11, 13, 17, 19, 23

9. Find the common factors of each set of numbers.
   (a) 21 and 42   (b) 48 and 60   (c) 25 and 100
   1, 3, 7, 21   1, 2, 3, 4, 6, 12   1, 5, 25

10. Find the first two common multiples for each set of numbers.
    (a) 1 and 7   (b) 3 and 9
    7, 14   9, 18
    (c) 3, 5, and 6   (d) 2, 5, and 10
    30, 60   10, 20

11. (a) If a number has 9 as a factor, is 3 also a factor of that number?
    Yes
    (b) If a number is a common multiple of 4 and 5, is 2 a factor of that number?
    Yes

3-6 Practice   77

---

90   Teacher's Guide 4A Chapter 3   © 2019 Singapore Math Inc.

Exercise 6 • page 63

Brain Works

▲ Fruit Baskets

The local food bank is assembling fruit baskets for donation. They receive fruit packed in large crates and divide them into the greatest number of identical baskets they can make with no fruit left over.

How many baskets can be made if the crates have:

- 16 apples and 12 oranges? 4
- 27 apples and 18 oranges? 9
- 21 apples and 35 oranges? 7
- 24 apples and 32 oranges? 8

★ Challenge:

- 16 apples, 8 oranges, and 12 bananas? 4
- 9 apples, 18 oranges, and 6 bananas? 3
- 27 apples, 45 oranges, and 63 bananas? 9
- 28 apples, 35 oranges, and 21 bananas? 7

---

12 Find the missing factors.

(a) $90 = 2 \times \boxed{45}$
$90 = 2 \times 3 \times \boxed{15}$
$90 = 2 \times 3 \times 3 \times \boxed{5}$

(b) $81 = 3 \times \boxed{27}$
$81 = 3 \times 3 \times \boxed{9}$
$81 = 3 \times 3 \times 3 \times \boxed{3}$

13 A bus for Route A leaves the station every 6 minutes. A bus for Route B leaves the same station every 9 minutes. Buses for both routes leave the station at 8:00 a.m. What is the next time that buses for both routes leave the station at the same time?
The first common multiple of 6 and 9 is 18.
8:18 a.m.

14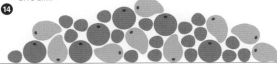

Clara is putting 9 mangoes, 18 lychees, and 6 oranges into baskets. Each basket must have the same combination of fruits. What is the greatest number of baskets she could have, and how many of each kind of fruit will be in each basket?
Common factors of 6, 9, and 18 are 1 and 3.
3 baskets of 3 mangoes, 6 lychees, and 2 oranges.

15 A number less than 70 is a multiple of 7. When it is divided by 6, there is a remainder of 1. When it is divided by 4, there is also a remainder of 1. What is the number? Multiples of 7: 7, 14, 21, 28, 35, 42, 49, 56, 63
7 and 49 have a remainder of 1 when divided by 6.
Of these two numbers, only 49 has a remainder of 1 when divided by 4.
49

Exercise 6 • page 63

78   3-6 Practice

Chapter 3 Multiples and Factors

Exercise 1

Basics

① (a) Find the first ten multiples of 7.

7, 14, 21, 28, 35, 42, 49, 56, 63, 70

(b) What is the 13th multiple of 7?
91

(c) What is the 7th multiple of 20?
140

(d) Is 105 a multiple of 7?
Yes, 105 ÷ 7 is 15 with no remainder.

② Without dividing, how can you easily determine if a number is...

(a) a multiple of 2?
The ones digit is 0, 2, 4, 6, or 8.

(b) a multiple of 3?
The sum of the digits is a multiple of 3.

(c) a multiple of 5?
The ones digit is 0 or 5.

③ Is 105 a multiple of 2, 3, or 5?
105 is a multiple of 3 and 5, but not 2.

Practice

④ (a) List the first twelve multiples of 5.
5, 10, 15, 20, 25, 30, 35, 40, 45, 50, 55, 60

(b) List the first four multiples of 17.
17, 34, 51, 68

⑤ Circle the numbers that are multiples of 6.

| 6 | 16 | 34 | 54 | 56 | 88 | 96 |

⑥ Circle the numbers that are multiples of 8.

| 6 | 16 | 34 | 54 | 56 | 88 | 96 |

⑦ (a) What is the 12th multiple of 9?
108

(b) What is the 11th multiple of 10?
110

(c) Is 110 a multiple of 4?
No. 110 ÷ 4 is 27 R 2.

(d) Is 136 a multiple of 4?
Yes. 136 ÷ 4 is 34 with no remainder.

⑧ | 15 | 36 | 60 | 64 | 78 | 120 | 216 |

Which of the above numbers are multiples of:

(a) 2?
36, 60, 64, 78, 120, 216

(b) 9?
36, 216

(c) 3?
15, 36, 60, 78, 120, 216

(d) 6?
36, 60, 78, 120, 216

(e) 5?
15, 60, 120

(f) 10?
60, 120

Challenge

⑨ A certain number is a multiple of 8. When 4 is added to it, the sum is the 4th multiple of 13. What is the number?
4th multiple of 13 = 52
52 − 4 = 48
48

⑩ In the pattern below, what shape will be in the 32nd position?

■▲●■▲●■▲●■▲●

The pattern repeats every 3 shapes. 32 ÷ 3 is 10 with a remainder of 2. The second shape is a triangle, so the 32nd shape is a triangle.

Exercise 2 • pages 50–52

Exercise 2

Basics

1 List the first twelve multiples of each number. Then list the common multiples.

(a) Multiples of 3:
3, 6, 9, 12, 15, 18, 21, 24, 27, 30, 33, 36
Multiples of 4:
4, 8, 12, 16, 20, 24, 28, 32, 36, 40, 44, 48
First three common multiples of 3 and 4:
12, 24, 36

(b) Multiples of 6:
6, 12, 18, 24, 30, 36, 42, 48, 54, 60, 66, 72
Multiples of 9:
9, 18, 27, 36, 45, 54, 63, 72, 81, 90, 99, 108
First three common multiples of 6 and 9:
18, 36, 54

2 List the first eight multiples of 8.
8, 16, 24, 32, 40, 48, 56, 64
Which of these are also multiples of 6?
24, 48
What are the first two common multiples of 6 and 8?
24, 48

3 21 is a common multiple of __3__ and __7__.

Is 42 also a common multiple of those two numbers?
Yes
Give the next two common multiples of those two numbers.
Multiples of 21 are common multiples of 3 and 7.
21 × 3 = 63, 21 × 4 = 84
63, 84

Practice

4 Find the first two common multiples of 8 and 12.
Multiples of 12: 12, 24, 36, 48
24 and 48 are also multiples of 8.
Common multiples: 24, 48

5 Circle the numbers that are common multiples of 3 and 4.

| 4 | **12** | 28 | 32 | **48** | 72 |

6 Find the first three common multiples of 2, 3, and 4.
Multiples of 4: 4, 8, 12, 16, 20, 24, 28, 32, 36
All multiples of 4 are multiples of 2.
12, 24, and 36 are also multiples of 3.
Common multiples: 12, 24, 36

7 Find the first three common multiples of 3, 5, and 6.
Multiples of 6: 6, 12, 18, 24, 30, ...
All multiples of 6 are multiples of 3.
30 is the first common multiple of 6 and 5, so the next two are 60 and 90
Common multiples: 30, 60, 90

8 Three numbers have a common multiple of 24. What is another common multiple of those three numbers?
Any multiple of 24 is a common multiple of the three numbers.
Example: 48

9 There are between 50 and 100 blocks. The blocks can be stacked equally into either 7 stacks or 5 stacks. How many blocks are there?
Multiples of 5: Any number that ends in 5 or 0.
The first common multiple of 5 and 7 is 35. The next ones will be 70 and 105. 70 is between 50 and 100.
There are 70 blocks.

10 3 bells ring at intervals of 3, 6, and 8 minutes. If they all rang together at 3:00, what time will they ring together again?
Multiples of 8: 8, 16, 24
24 is the first one that is a multiple of 6.
All multiples of 6 are multiples of 3.
24 minutes after 3:00 is 3:24.
3:24

Challenge

11 A number, when divided by 5 gives a remainder of 3, and when divided by 4 gives a remainder of 1. What is the least number it could be?
Multiples of 5, added to 3: 8, 13, 18, 23, 28, 33
Multiples of 4, added to 1: 5, 9, 13, 17, 21, 25, 29, 33
13

12 A number, X, is a common multiple of 2, 6, and 8. Another number, Y, is a common multiple of 3, 5, and 10. What is the least possible value for X + Y?
Multiples of 8: 8, 16, 24
24 is also a multiple of 6 and 2.
Multiples of 10: 10, 20, 30
30 is also a multiple of 3 and 5.
24 + 30 = 54

Exercise 3 • pages 53–56

Exercise 3

Basics

1 Find the factors of 16.

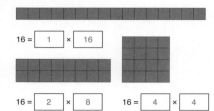

16 = [1] × [16]

16 = [2] × [8]   16 = [4] × [4]

The factors of 16 are __1__, __2__, __4__, __8__, and __16__.

2 (a) Find the factors of 35.

35 = 1 × [35]   35 = 5 × [7]   35 = 7 × [5]

The factors of 35 are __1__, __5__, __7__, and __35__.

(b) How can we tell without dividing that 2, 4, and 6 are not factors of 35?
2, 4, and 6 are all even numbers and 35 is odd.

3 (a) 8)9 4 = 1 1 R 6

Is 8 a factor of 94? No

(b) 8)9 6 = 1 2

Is 8 a factor of 96? Yes

Practice

4 Write "yes" or "no."

Number	Is 2 a factor?	Is 3 a factor?	Is 5 a factor?
5	No	No	Yes
15	No	Yes	Yes
36	Yes	Yes	No
60	Yes	Yes	Yes
73	No	No	No
84	Yes	Yes	No
100	Yes	No	Yes
114	Yes	Yes	No
120	Yes	Yes	Yes

5 (a) Circle the numbers that are factors of 18.

1 2 3 4 5 6 7

(b) Circle the numbers that are factors of 48.

3 5 6 8 10 48 96

(c) Circle the numbers that have 6 as a factor.

1 3 6 26 30 72 100

6 Find all the factors of each of the following numbers. List them in order from least to greatest.

(a) 56
1, 2, 4, 7, 8, 14, 28, 56

(b) 64
1, 2, 4, 8, 16, 32, 64

(c) 80
1, 2, 4, 5, 8, 10, 16, 20, 40, 80

(d) 96
1, 2, 3, 4, 6, 8, 12, 16, 24, 32, 48, 96

(e) 120
1, 2, 3, 4, 5, 6, 8, 10, 12, 15, 20, 24, 30, 40, 60, 120

7 32 sandwiches are arranged on plates so each plate gets the same number of sandwiches. What are the possible numbers of plates needed?
1, 2, 4, 8, 16, 32

8 54 flowers are to be arranged in an even number of vases so that each vase has the same number of flowers. There needs to be at least 5 vases and at most 20 vases. What are the possible numbers of vases needed?
Factors of 54: 1, 2, 3, 6, 9, 18, 27, 54
6 or 18 vases

Challenge

9 Circle the numbers that have an odd number of factors.

9 10 15 25 49 81 121

10 What are the four numbers less than 50 that have exactly 3 factors?
Numbers with an odd number of factors are 1 × 1, 2 × 2, 3 × 3, etc, so check 1, 4, 9, 16, 25, 36, and 49.
The ones with 3 factors are 4, 9, 25, and 49.

Exercise 4 • pages 57–59

Exercise 4

Basics

1 (a) Find all the factors of each of the following numbers.

19	20
1, 19	1, 2, 4, 5, 10, 20
21	23
1, 3, 7, 21	1, 23
27	29
1, 3, 9, 27	1, 29

(b) Which of these numbers have only 2 factors, and so are prime numbers?
19, 23, 29

(c) Which of these numbers have more than 2 factors, and so are composite numbers?
20, 21, 27

2 What two numbers are neither prime nor composite?
0, 1

Practice

3 (a) The prime numbers less than 10 are __2__, __3__, __5__, and __7__.

(b) The only even number that is a prime number is __2__.

(c) Other than 5, which is prime, all odd numbers with __5__ in the ones place are composite numbers since they are multiples of 5.

(d) Which of the following remaining odd numbers less than 100 are composite numbers? Cross them off.

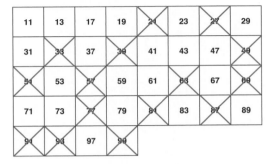

(e) Using the above information, list the 25 prime numbers less than 100.
2, 3, 5, 7, 11, 13, 17, 19, 23, 29, 31, 37, 41, 43, 47, 53, 59, 61, 67, 71, 73, 79, 83, 89, 97

4 Twin primes are pairs of prime numbers that have a difference of 2, for example, 11 and 13. Find all the twin primes less than 100.
3 and 5, 5 and 7, 11 and 13, 17 and 19, 29 and 31, 41 and 43, 59 and 61, 71 and 73

5 16 can be expressed as the sum of two prime numbers: 16 = 11 + 5. Find three other numbers greater than 10 and show how they can each be expressed as the sum of two prime numbers.
Answers will vary. Any even number can be expressed as the sum of two prime numbers.

6 An emirp is a pair of prime numbers with reversed digits, such as 37 and 73. List all the pairs of two-digit emirps.
Since numbers where the tens digit is even will be even numbers if the digits are reversed, and not prime, only numbers where the digit in the tens place is odd need to be checked.
13 and 31, 17 and 71, 37 and 73, 79 and 97.

7 Circle the number in the grid that obeys the following two rules.

Rule 1 The number is not in a row that contains a prime number.
Rule 2 The number is not in a column that contains a number that has an odd number of factors.

82	27	12	31
10	19	42	36
81	40	28	18
79	25	65	34

(28 is circled)

Exercise 5 • pages 60–62

Exercise 5

Basics

1. (a) 12 = 1 × [12]
   12 = 2 × [6]
   12 = 3 × [4]

   The factors of 12 are __1__, __2__, __3__, __4__, __6__, and __12__.

   (b) 20 = 1 × [20]
   20 = 2 × [10]
   20 = 4 × [5]

   The factors of 20 are __1__, __2__, __4__, __5__, __10__, and __20__.

   (c) The common factors of 12 and 20 are __1__, __2__, and __4__.

   (d) Which common factor of 12 and 20 is greatest? __4__

2. Circle common factors of 36 and 42.

   [2] [3] [6] [9] [12] [18] [36]

Practice

3. (a) Circle the numbers that have 4 and 5 as common factors.

   [5] [20] [35] [40] [48] [65] [100]

   (b) Circle the numbers that have 2, 3, and 6 as common factors.

   [6] [18] [24] [30] [42] [81] [120]

4. Find all the common factors of each set of numbers.

   (a) 24, 30
   1, 2, 3, 6

   (b) 48, 84
   1, 2, 3, 4, 6, 12

   (c) 35, 37, 90
   1

5. 75 black pens, 45 blue pens, and 30 red pens need to be put in boxes so that each box has the same number of each color of pen. What is the greatest number of boxes needed, and how many of each color pen will be in each box?

   Common factors of 75, 45, and 30: 1, 3, 5, 15
   Greatest common factor: 15

   Greatest number of boxes: 15
   Black pens: 75 ÷ 15 = 5
   Blue pens: 45 ÷ 15 = 3
   Red pens: 30 ÷ 15 = 2
   15 boxes; 5 black pens, 3 blue pens, 2 red pens

Challenge

6. Continue a path through the maze. In order to move from one square to an adjacent square, the two numbers must have a common factor other than 1.

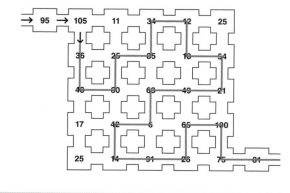

Exercise 6 • pages 63–66

Exercise 6

Check

1. List the multiples of 6 that are between 30 and 80.
   36, 42, 48, 54, 60, 66, 72, 78

2. What number less than 200 is the greatest multiple of 5?
   195

3. Find the sum of the first four multiples of 3 and the first three multiples of 4.
   Multiples of 3: 3, 6, 9, 12
   Multiples of 4: 4, 8, 12
   3 + 6 + 9 + 12 + 4 + 8 + 12 = 54

4. Find the first three common multiples of each set of numbers.
   (a) 6, 15
       30, 60, 90
   (b) 8, 10, 12
       120, 240, 360

5. (a) What number is the least multiple of 38?
       38
   (b) What number is the greatest possible factor of 120?
       120

6. A 2-digit even number has 5 and 9 as factors. What number is it?
   Common multiples of 5 and 9: 45, 90, 135
   90

7. A 2-digit odd number is a factor of 54 and a multiple of 9. What number is it?
   Factors of 54: 1, 2, 3, 6, 9, 18, 27, 54
   Multiples of 9: 9, 18, 27, 36, 45, 54
   27

8. Find all the common factors of each set of numbers.
   (a) 30, 45
       1, 3, 5, 15
   (b) 30, 40, 100
       1, 2, 5, 10

9. Can the product of two prime numbers be an even number? Explain why or why not.
   Yes, if one of the prime numbers is 2. The product of any other pair of prime numbers would be odd, since the product of two odd numbers is odd.

10. Can the product of two prime numbers be a prime number? Explain why or why not.
    No. If a number is the product of two numbers (other than 1), then it is not a prime number.

11. A blue light blinks every 10 seconds. A red light blinks every 12 seconds. A green light blinks every 8 seconds. If all three lights just blinked at the same time, in how many seconds will they again blink at the same time?
    Least common multiple of 8, 10, and 12: 120
    120 seconds

12. A red ribbon is 64 inches long, a blue ribbon is 80 inches long, and a yellow ribbon is 96 inches long. They are all to be cut into shorter pieces of the same length without any leftover pieces. What is the longest possible length of each ribbon? How many pieces of each color of ribbon will there be?
    Common factors of 64, 80, and 96: 1, 2, 4, 8, 16
    Longest possible length is 16 cm.
    Red ribbon: 64 ÷ 16 = 4; 4 pieces
    Blue ribbon: 80 ÷ 16 = 5; 5 pieces
    Yellow ribbon: 96 ÷ 16 = 6; 6 pieces

Challenge

13. A group of people are traveling together to camp. Each car can hold 5 people. If 4 people were to go in each car, there would be 3 people left over. If 5 people were to go in each car, there would be room for 2 more people in the last car. How many people and how many cars are there?
    When there are 4 people in each car, the remainder is 3. When there are 5 people in each car, the remainder is also 3, since 2 seats will be empty, leaving 3 occupied seats.
    Multiples of 4, added to 3: 7, 11, 15, 19, 23
    Multiples of 5, added to 3: 8, 13, 18, 23
    4 people in one car: 23 ÷ 4 is 5 R 3; 5 people in one car: 23 ÷ 5 is 4 R 3
    23 people and 5 cars

14. Emma found the difference between two prime numbers is 15. What are the two prime numbers?
    2, 17
    All prime numbers except for 2 are odd. The difference between two odd numbers is an even number. So one of the numbers must be 2.

15. A Sophie Germain prime is a prime number such that if 1 is added to twice that number, the answer is also a prime number. For example, 3 is a Sophie Germain prime number because 2 × 3 + 1 = 7, which is also prime. Find all the Sophie Germain primes that are less than 50.
    2, 3, 5, 11, 23, 29, and 41

Notes

Chapter 4 Multiplication — Overview

Suggested number of class periods: 8–9

	Lesson	Page	Resources	Objectives
	Chapter Opener	p. 105	TB: p. 79	Review multiplication concepts.
1	Mental Math for Multiplication	p. 106	TB: p. 80 WB: p. 67	Multiply up to a four-digit number by a one-digit number using mental math.
2	Multiplying by a 1-Digit Number — Part 1	p. 110	TB: p. 84 WB: p. 69	Multiply up to a four-digit number by a one-digit number.
3	Multiplying by a 1-Digit Number — Part 2	p. 113	TB: p. 88 WB: p. 73	Multiply up to a four-digit number by a one-digit number.
4	Practice A	p. 116	TB: p. 91 WB: p. 77	Practice multiplication.
5	Multiplying by a Multiple of 10	p. 119	TB: p. 94 WB: p. 80	Multiply a two-digit number by a two-digit multiple of 10.
6	Multiplying by a 2-Digit Number — Part 1	p. 123	TB: p. 98 WB: p. 82	Multiply a two-digit number by a two-digit number using the standard algorithm.
7	Multiplying by a 2-Digit Number — Part 2	p. 125	TB: p. 101 WB: p. 86	Multiply a three-digit number by a two-digit number.
8	Practice B	p. 127	TB: p. 104 WB: p. 90	Practice multiplication.
	Workbook Solutions	p. 130		

Chapter 4 Multiplication

Notes

In Dimensions Math 3A students learned how to use the standard algorithm to multiply up to a three-digit number by a one-digit number.

In this chapter, students will extend their understanding of the algorithm to four-digit numbers. They will then learn to multiply four-digit numbers by two-digit numbers.

Mental Math for Multiplication

In Lesson 1, students extend place value concepts learned in Chapter 1 Lesson 7 to mental math calculations for multiplication, which will be used to estimate products.

Students will see multiples of tens, hundreds, and thousands that can easily be solved by thinking of simple two-digit and one-digit computations. In the problem 4,200 × 3, students can begin by thinking of 42 × 3. They will multiply by splitting 42 into 40 and 2, multiplying each number by 3, and adding the two products.

$$42 \times 3 = 120 + 6$$
$$\diagup \quad \diagdown$$
$$40 \qquad 2$$

With their knowledge of place value, they can then find the product of 4,200 and 3 by keeping in mind that 42 ones × 3 = 126 ones, so 42 hundreds × 3 = 126 hundreds.

42 hundreds × 3 = 126 hundreds or 12,600

This method explains the common shorthand strategy of "appending zeros." Students are computing the calculation of the value in the hundreds place, and then multiplying that product by 100.

To solve multiplication problems with numbers near a multiple of 100, students will learn to multiply by the multiple of 100 and subtract the difference.

For example, 299 × 4 can be thought of as (300 × 4) − (1 × 4) = 1,200 − 4, or 1,196.

Standard Multiplication Algorithm

Below is a general procedure for demonstrating the multiplication algorithm with place-value discs. Students should be comfortable with all the steps, including regrouping.

For a more basic review of the multiplication algorithm, reference Dimensions Math 3A Chapter 5.

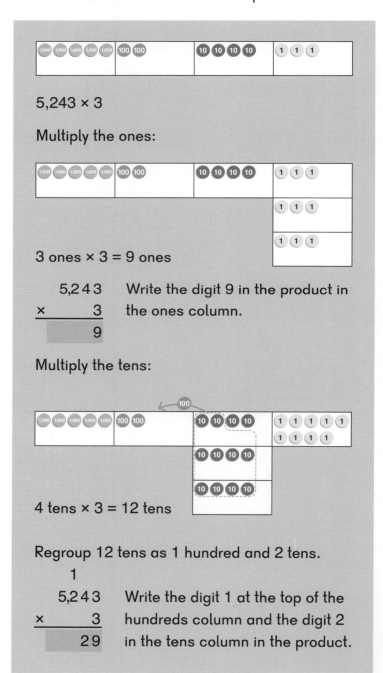

Chapter 4 Multiplication

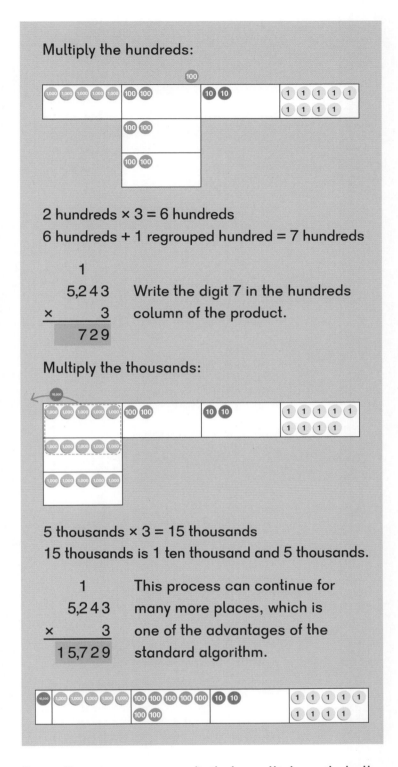

Multiply the hundreds:

2 hundreds × 3 = 6 hundreds
6 hundreds + 1 regrouped hundred = 7 hundreds

```
   1
 5,243    Write the digit 7 in the hundreds
×    3    column of the product.
   729
```

Multiply the thousands:

5 thousands × 3 = 15 thousands
15 thousands is 1 ten thousand and 5 thousands.

```
   1
 5,243    This process can continue for
×    3    many more places, which is
15,729    one of the advantages of the
          standard algorithm.
```

Since there are no more digits to multiply, write both the 1 ten thousand and 5 thousands in the correct place value columns.

Multiplying by Two-digit Numbers

Students should have sufficient experience estimating products before multiplying by two-digit numbers.

Estimating products helps students check for errors in place value, which is important when multiplying by two-digit numbers.

To multiply by a two-digit number, students will first apply the distributive property to multiples of ten. For example, with the problem 34 × 20, students will learn that this product has the same value as 34 × 10 × 2.

Next, students will learn how to multiply up to a three-digit number by a two-digit number that has two non-zero digits.

Students begin by informally using the distributive property. For example, 34 × 12 = (34 × 10) + (34 × 2).

Students will then see the same problem in a vertical format in which the number being multiplied is first multiplied by the ones. The product is written, and then multiplied by the tens. Then the final product is written.

For 34 × 12, students will first multiply 34 by 2 and then by 10.

```
    34
×   12
    68   ← 34 × 2
   340   ← 34 × 10
```

34 × 1 **ten** is 34 **tens**, or 34**0**. Students are encouraged to write the 0 in the ones column to remember they are multiplying tens. This will also help to line up the numbers according to place value.

While it is possible to multiply by tens first and then by ones, the products will get larger as lessons progress and students should be encouraged to start calculating in the smallest place.

Chapter 4 Multiplication

For textbook multiplication problems that require regrouping, the regrouped numbers are shown above the number being multiplied:

We can write the regrouped digits on top.

```
    4 4
    6 6
    4 8 9
  ×   5 7
  ───────
    3,4 2 3   ← 489 × 7
      4 5 0   ← 489 × 50
```

Students can cross off the regrouping marks after they have completed multiplying by the ones, and record the regrouping for multiplying by the tens above the crossed off numbers.

```
    4 4
    6̶ 6̶
    4 8 9
  ×   5 7
  ───────
    3,4 2 3
      4 5 0
```

Students who can work mentally with the regrouped numbers do not need to write the regrouping marks.

Students who need additional practice with lining up numbers by the correct place value can practice either on graph paper or wide-ruled notebook paper turned sideways, so the lines make columns.

Area Models

An area model is a way to see the partial products and can be shown to students if needed. Students calculate the area of smaller portions of the entire rectangle and add them together to find the area of the whole shape.

For example, when multiplying 489 by 57:

	400	80	9
50	400 × 50 = 20,000	80 × 50 = 4,000	9 × 50 = 450
7	400 × 7 = 2,800	80 × 7 = 560	9 × 7 = 63

```
         4 8 9
      ×    5 7
      ────────
           6 3   ← 9 × 7     ⎫
         5 6 0   ← 80 × 7    ⎬ 489 × 7
       2,8 0 0   ← 400 × 7   ⎭
         4 5 0   ← 9 × 50    ⎫
       4,0 0 0   ← 80 × 50   ⎬ 489 × 50
      2 0,0 0 0  ← 400 × 50  ⎭
      ────────
      2 7,8 7 3
```

Using an area model and partial products for large multi-digit computations is cumbersome, and students should master the standard algorithm.

Bar Models

Students should have prior experience using bar models to help them solve word problems. If students are unfamiliar with bar models, reference Dimensions Math 3A Chapter 4 Lessons 7–9.

For reference, below is a review of the two main types of bar models students learned in Dimensions Math 3.

Part-whole Models

This type of model extends the understanding of multiplicative relationships. Each part is an equal group and is considered a "unit."

Chapter 4 Multiplication

There are 2,300 nails in 1 box. How many nails are in 5 boxes?

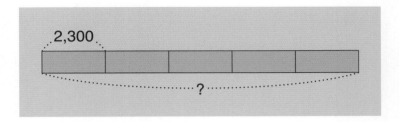

1 unit ⟶ 2,300
5 units ⟶ 5 × 2,300 = 11,500
There are 11,500 nails in 5 boxes.

Comparison Models

This type of bar model is always used when comparing quantities and is particularly useful in representing multi-step problems.

There are 750 sheets of paper in a small box. A large box contains 3 times as many sheets of paper as a small box. How many sheets of paper are in the large box?

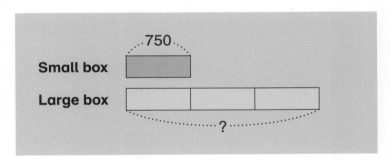

1 unit ⟶ 750
3 units ⟶ 3 × 750 = 2,250
There are 2,250 sheets of paper in the large box.

This same model will help students solve these questions:

(a) What is the total number of sheets of paper in both boxes?

(b) How many more sheets of paper are in the large box than in the small box?

There are 750 sheets of paper in a small box. A large box contains 3 times as many sheets of paper as a small box. How many sheets of paper are there in all?

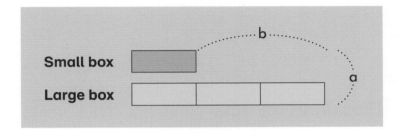

(a) 1 unit ⟶ 750
4 units ⟶ 4 × 750 = 3,000
There are 3,000 sheets of paper in both boxes.

(b) 1 unit ⟶ 750
2 units ⟶ 2 × 750 = 1,500
1,500 more sheets of paper are in the large box than in the small box.

Bar models for multiplication and division use equal units. While drawing models, students should be asked questions such as:

- "Which bar represents the unit?"
- "How many units are needed? How do you know?"
- "What is known in the problem? What must we find?"
- "What is unknown in the problem? How do we know?"

Chapter 4 Multiplication

Materials

- 10-sided die
- Deck of cards with face cards removed
- Dry erase markers
- Place-value discs
- Place-value organizer
- Whiteboards

Blackline Masters

- Number Cards
- Mental Math Duel Scoring Sheet

Activities

Activities included in this chapter provide practice with multiplication and division. They can be used after students complete the **Do** questions, or anytime additional practice is needed.

Chapter Opener

Objective

- Review multiplication concepts.

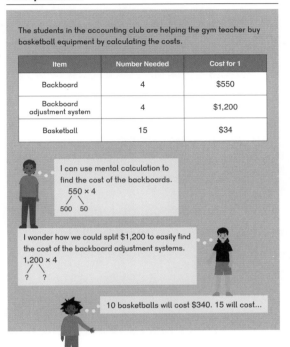

Have students discuss the mental math strategies in the **Chapter Opener**. They may find different ways to calculate the costs of the equipment.

Alex uses a strategy from Dimensions Math 3A. He splits 550 into 500 and 50 and multiplies each number by 4.

Mei also splits 1,200 into two parts, then finds the product of each part when multiplied by 4. She could split 1,200 in multiple ways.

For example:

- 1,000 × 4 and 200 × 4
- 600 × 4 and 600 × 4 (or 600 × 2 × 4)

Dion knows that 5 basketballs will be the same as half of the cost of 10 basketballs. He could split 340 in multiple ways.

For example:

- Divide 340 by 2, and add that quotient to 340 to find the cost of 15 basketballs: 340 + 170 = 510.
- Divide 340 by 2 and multiply the quotient by 3: 170 × 3 = 100 × 3 + 70 × 3 = 300 + 210 = 510.

The **Chapter Opener** can continue straight to Lesson 1.

Lesson 1 Mental Math for Multiplication

Objective

- Multiply up to a four-digit number by a one-digit number using mental math.

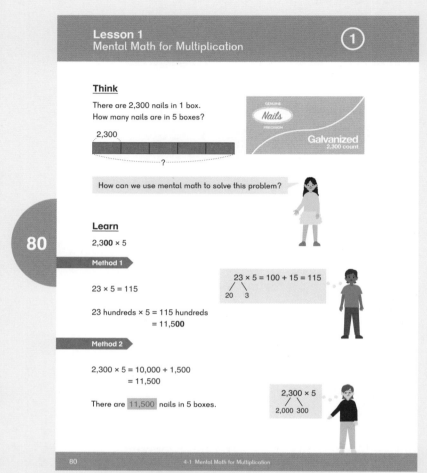

Think

Pose the **Think** problem and ask students to use a mental math method to find the number of nails in five boxes. If students finish quickly, challenge them to solve the problem in more than one way.

Discuss student strategies.

This is the first multiplication model used in the textbook. Ask students about the bar model shown in **Think**.

- "What information is given?"
- "How is the given information shown on the model?"
- "What is the question mark asking us to find?" (The product or the whole.)
- "What can we say about the individual boxes in the model?" (They are all equal, or they all represent the same amount.)

Learn

Discuss the methods shown in **Learn** and have students compare their own methods with the methods shown in the textbook.

Alex thinks of 2,**300** as 23 **hundreds** and multiplies 23 by 5 by splitting 23 into 20 and 3. He then appends two zeros to that product for hundreds.

Emma splits 2,300 into thousands and hundreds and multiplies them each by 5. She then adds the partial products together to get 11,500.

106 Teacher's Guide 4A Chapter 4 © 2019 Singapore Math Inc.

Do

1—**2** Dion and Mei think about using number bonds to calculate mentally. Students can use this method combined with their knowledge of place value to easily see the patterns and relationships. 32 × 4 is the same computation whether the number being multiplied by 4 is 32 ones, 32 tens, 32 hundreds, or 32 thousands.

32 hundreds × 4 = 128 hundreds
32 thousands × 4 = 128 thousands

Students should also see that once they find 32 × 4, they can append one zero to find 32 tens × 4 or two zeros for 32 hundreds × 4.

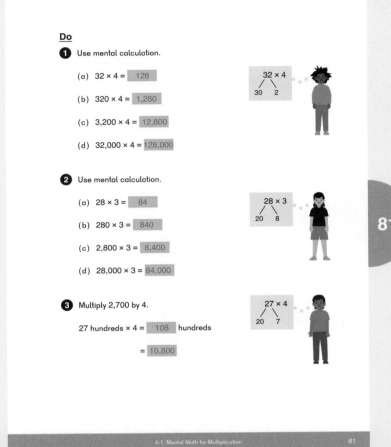

5. Dion introduces a strategy for numbers close to 100. 100 × 8 is 8 more than 99 × 8.

Students who struggle to see why this works should be encouraged to recall single digit facts: 9 × 8 was 9 less than 10 × 8. Use arrays if necessary to show the relationship.

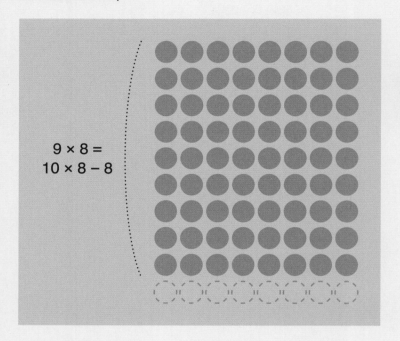

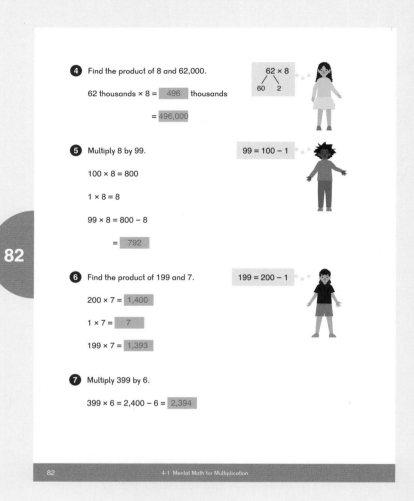

Students can then apply that knowledge to:

99 × 8 = (100 − 1) × 8
       = (100 × 8) − (1 × 8)
       = 800 − 8
       = 792

6. Mei notes the same strategy for multiples of 100: 200 × 7 is 7 more than 199 × 7.

7. Students should see that 399 × 6 = 400 × 6 − 6.

108   Teacher's Guide 4A Chapter 4   © 2019 Singapore Math Inc.

8 Have students share their mental math methods for solving some of the problems.

9 Ask students about the bar model shown and compare it to the model in **Think**.

- "Why are there 2 different bars here?"
- "What information is given?"
- "How is the given information shown on the model?"
- "What is the question mark asking us to find?"
- "What can we say about the individual boxes in the model?"
- "What other information could we find using this model? How many more sheets of paper are in the large box than the small box?" (750 × 2)
- "How many sheets of paper are in both boxes?" (750 × 4)

Activity

▲ Mental Math Battle

Materials: Deck of cards with face cards removed

In each round, one card from the deck is placed faceup in the middle. That card is the number by which each player's two-digit number will be multiplied.

On each round, deal two cards to each player. Players use their cards to make a two-digit number and use mental math to multiply their two-digit number by the card already turned over.

The player with the greatest product gets a point. The winner is the player with the most points at the end of the game or allotted time.

★ Extend the activity by having Player One deal each player 3 cards, facedown. Players choose two of their cards to look at and create a two-digit number. Player One then calls "1, 2, 3." On 3, all players turn over their third card and multiply that number by the two-digit number they have in their hands. The first player to call the correct product takes all the cards. The winner is the player with the most cards after five rounds.

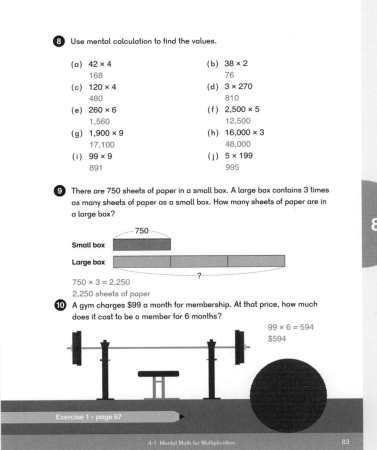

Exercise 1 • page 67

Lesson 2 Multiplying by a 1-Digit Number — Part 1

Objective

- Multiply up to a four-digit number by a one-digit number.

Lesson Materials

- Place-value discs
- Place-value organizer

Think

Provide students with place-value discs and pose the **Think** problem. Have students estimate the cost first, then find an actual cost. They can use place-value discs, the algorithm, or any other known strategy to find the product.

Discuss student estimates and ideas for finding the actual amount.

Learn

Discuss the methods shown in **Learn** and have students compare their own methods with the methods shown in the textbook.

Alex rounded 2,415 down to 2,000 and then multiplied by 3.

Dion rounded to 2,500. His estimate is greater than the actual product.

Ask students why they think one estimate is more accurate than the other.

Students could also have estimated the product using: 2,400 × 3, which is 6,000 + 1,200.

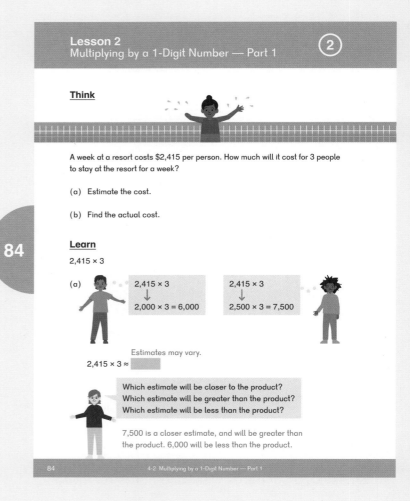

110   Teacher's Guide 4A Chapter 4   © 2019 Singapore Math Inc.

Students should recall the standard algorithm for multiplying a three-digit number by a one-digit number from Dimensions Math 3. Discuss how the algorithm is similar for four-digit numbers.

Have students relate the steps in the algorithm as shown with the place-value discs in the textbook.

Using their knowledge of place value, they know that when there are more than 10 of a place value they need to regroup 10 of them to the next higher place value.

During **Think**, students may also have used a partial product algorithm to find the actual amount:

```
    2,4 1 5
  ×       3
         15
         30
      1,200
      6,000
      7,245
```

If students used the partial product algorithm, point out how the numbers in the partial product algorithms correspond to the numbers using the traditional algorithm. For example, the 1 ten in 15 is the 1 that was regrouped to the tens place.

Mei asks if the answer is reasonable. Students can compare their answers to their estimates to see if they are close.

Do

❶ Estimates should be quick. Most students will use 9,000 × 4 = 36,000. Because 9,000 is more than 8,712, students should see that this estimate will be greater than the actual product.

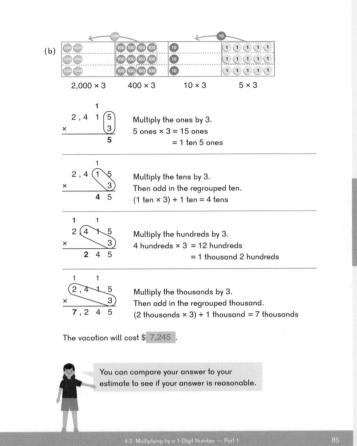

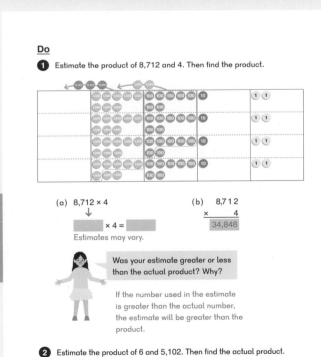

③ Have students share their estimates for some of the problems.

⑤ Discuss the bar model with students.

- "Why are there 2 different bars here?"
- "What information is given?"
- "How is the given information shown on the model?"
- "What is the question mark asking us to find?"
- "What can we say about the individual boxes in the model?"

Students may also solve the problem by finding the cost of the indoor pool first:

Outdoor pool: 1 unit ⟶ 6,242
Indoor pool: 3 units ⟶ 6,242 × 3 = 18,726

They can then add the cost of the two pools together:
6,242 + 18,726 = 24,968

Students who solve the problem this way can see from the given model in the textbook that it can be solved more efficiently.

Activity

▲ Mental Math Duel

Materials: Number Cards (BLM), multiple sets, or 10-sided die, Mental Math Duel Scoring Sheet (BLM)

Players take turns drawing a card (or rolling the die) and filling in the nine boxes with the numbers on the card or die. After drawing nine cards (or rolling the die nine times), players sheets should be filled out. Once players have filled out their scoring sheets, they next complete the multiplication equations and add the three products together.

The player with the greatest sum is the winner.

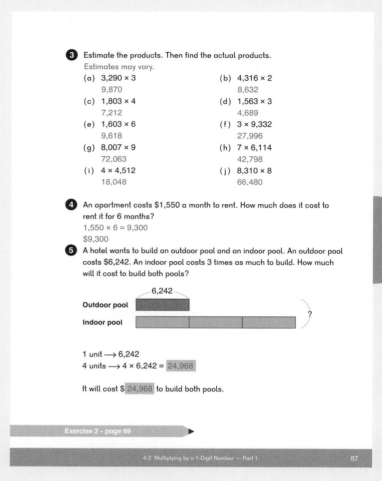

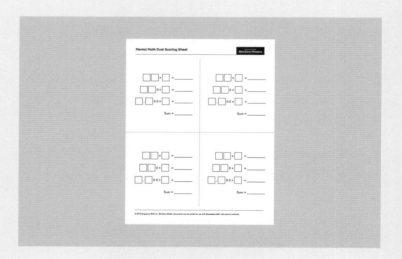

Exercise 2 • page 69

Lesson 3 Multiplying by a 1-Digit Number — Part 2

Objective

- Multiply up to a four-digit number by a one-digit number.

Lesson Materials

- Place-value discs
- Place-value organizer

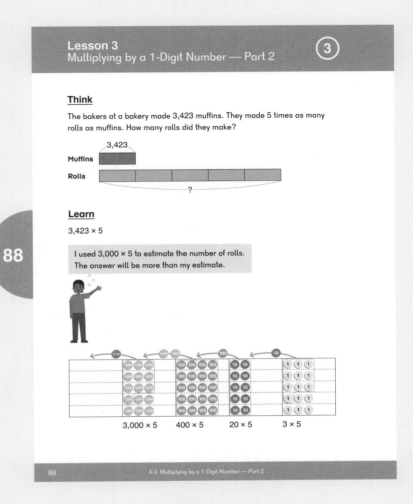

Think

Pose the **Think** problem. Have students estimate how many rolls the bakery made before finding the actual amount.

Discuss the model with students.

- "Why are there two different bars here?"
- "What information is given?"
- "How is the given information shown on the bar model?"

Discuss student solutions for the problem. Most students should not need the discs at this point and should solve the problem simply using the algorithm.

Learn

Students can see how the regrouping of the place-value discs corresponds to the regrouping in the problem solved using the algorithm.

Discuss the steps in the algorithm. Have students note where the regrouped amount is recorded when multiplying ones, then tens, then hundreds.

Students should note that when they have 17 thousands they do not need to record the regrouping, but can simply write the number of thousands (17) in the product.

During **Think**, students may also have used a partial product algorithm to find the actual amount:

```
    3,4 2 3
  ×       5
         15  ← 3 × 5
        100  ← 20 × 5
      2,000  ← 400 × 5
     15,000  ← 3,000 × 5
     17,115
```

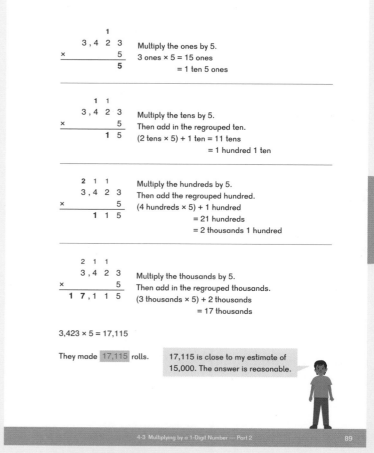

If students used a partial product algorithm, point out how the numbers in the partial product correspond to the numbers using the traditional algorithm. For example, the 1 ten in 15 is the 1 that was regrouped to the tens place.

Do

2 Have students share their estimates for some of the problems.

(g) Students may also solve using a mental math strategy: 10,000 × 9 − 1 × 9 = 90,000 − 9.

3 Discuss the model with students and ask:

- "Why are there two different bars here?"
- "What information is given?"
- "How is the given information shown on the model?"

Students may also solve the problem by finding the number of chicken eggs first:

Duck eggs: 1 unit ⟶ 3,489
Chicken eggs: 4 units ⟶ 3,489 × 4 = 13,956
13,956 − 3,489 = 10,467

Students who solve the problem this way can see from the model that the method shown in the textbook requires fewer steps to solve the problem.

Activity

▲ Greatest Product

Materials: Number Cards (BLM) 0–9, four sets, or deck of cards with face cards removed

Place the deck of cards facedown in the middle of the players (2–6 players). On each turn, a player draws five cards and uses four cards to create a four-digit number. They then multiply that four-digit number by the number on the remaining card.

When each player has had a turn, players compare their products. The player with the greatest product gets a point. The first player to get 10 points is the winner.

Exercise 3 • page 73

Lesson 4 Practice A

Objective

- Practice multiplication.

❷ Have students share their estimates for some of the problems.

❸ Students should estimate first. The estimate of 500 × 7 will be greater than the actual product of 460 × 7, however, in this case the estimate of 3,500 is less than the given cost so there must be a mistake.

To find the error, students can first calculate the actual amount: 460 × 7 = 2,800 + 420 = 3,220.

Students might say:

- "Mrs. Green wrote a 9 instead of a 2 in the hundreds place."
- "3,920 − 3,220 = 700. 700 divided by 7 = 100, so Mrs. Green multiplied 7 by 560 instead of 460."

Lesson 4 Practice A

❶ Multiply.

(a) 4 × 320
1,280

(b) 2,500 × 4
10,000

(c) 5 × 6,200
31,000

(d) 1,200 × 7
8,400

(e) 34,000 × 6
204,000

(f) 4 × 59
236

(g) 5 × 99
495

(h) 3 × 299
897

❷ Estimate the products. Then find the actual products.
Estimates may vary. Actual products provided.

(a) 375 × 4
1,500

(b) 6 × 1,072
6,432

(c) 6 × 289
1,734

(d) 4,309 × 7
30,163

(e) 5,876 × 5
29,380

(f) 9 × 5,845
52,605

(g) 16,479 × 2
32,958

(h) 3 × 34,145
102,435

❸ Ms. Green is buying 460 cups for a restaurant. Each cup costs $7. She estimated the cost by multiplying 500 by 7. When she calculated the actual cost as $3,920, she knew immediately that she made a mistake. How did she know? What error did she make?
Estimate: 500 × 7 = 3,500
The estimated product will be greater than the actual answer. $3,920 is even greater than the estimate. The clerk multiplied 560 by 7 instead of 460 by 7.

4. Ask students if using an estimate of more than 435 makes sense here. If they round to 400, they can see that Martin already meets his goal, which means that saving $435 a month will definitely meet his goal.

5. Discuss the problem with students. From the bar model given, they should see that they do not need to find the number of Jersey cows or Holstein cows to solve the problem.

   Ask students where the question marks would be placed for each problem.

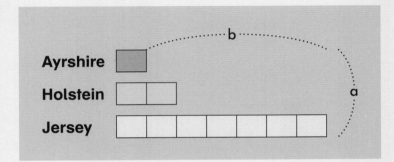

(a) If 1 unit represents 800 cows, and there are 10 total units:

1 unit ⟶ 800
10 units ⟶ 800 × 10 = 8,000
There are 8,000 cows in all.

(b) To find the difference between the number of Jersey cows and Ayrshire cows, students can count the units:

6 units ⟶ 800 × 6 = 4,800
There are 4,800 more Jersey cows.

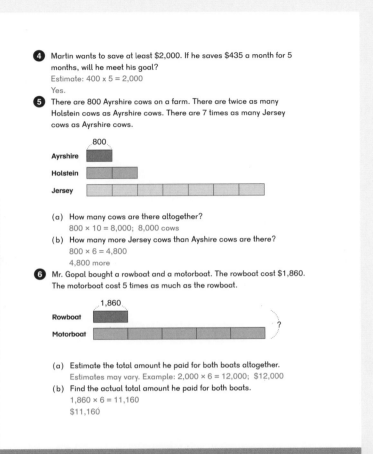

7 Discuss the multi-step problem with students.

Ask students:

- "What does the top bar represent?" (6 months of Ms. Chen's savings.)
- "What does the bottom bar represent?" (The cost of the motor scooter.)
- "What does the question mark in the bar model ask us to find?" (The amount of money Ms. Chen has left after buying the scooter.)

Ms. Chen savings:

1 unit ⟶ 445
6 units ⟶ 445 × 6 = 2,670

1,899 + ? = 2,670, or 2,670 − 1,899 = 771

8 Discuss the bar model with students.

Ask students:

- "Why are there before and after bars?"
- "What information is given in the problem and how is it represented on the model?"
- "What does the part of the after bar for cows with the dotted lines represent?"
- "How can we find the value of 1 unit?"

Students should see that the total number of goats does not change. We can find 1 unit by finding:

320 − 35 = 285

There were 3 total units of goats and cows at first:

1 unit ⟶ 285
3 units ⟶ 285 × 3 = 855

Exercise 4 • page 77

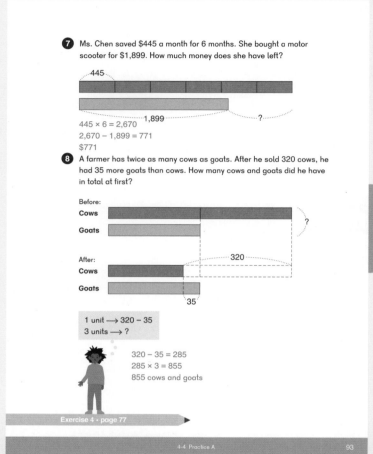

Lesson 5 Multiplying by a Multiple of 10

Objective

- Multiply a two-digit number by a two-digit multiple of 10.

Lesson Materials

- Place-value discs, ten each of the values 1, 10, 100

Think

Pose the **Think** problem. Students should think about how to multiply 34 by 2 tens. Have students think about how they could show 34 × 20 with only the discs provided. They do not have enough discs to make 20 groups of 34.

If students finish quickly, challenge them to solve the problem in more than one way.

Discuss student strategies.

Learn

Discuss the methods shown in **Learn** and have students compare their own methods with the methods shown in the textbook.

Sofia thinks of 20 as 10 × 2. She multiplies 34 by 10 first and then by 2.

To demonstrate if needed:

- 34 × 10 would be 30 ten discs and 40 one discs, which can be regrouped to 3 hundred discs and 4 ten discs.
- 3 hundred discs and 4 ten discs multiplied by 2 is 6 hundred discs and 8 ten discs.

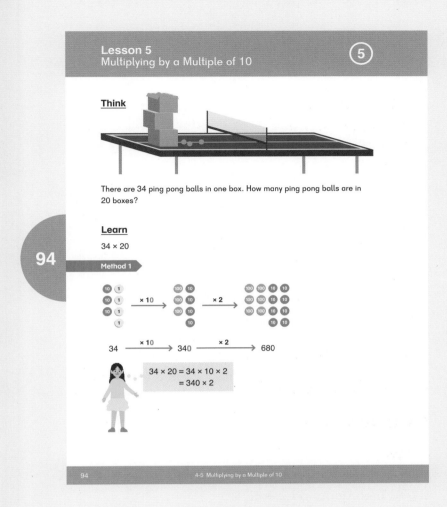

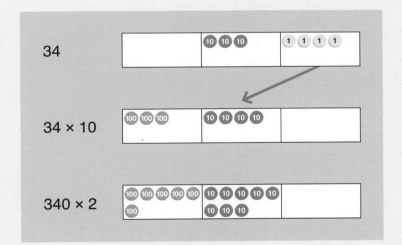

© 2019 Singapore Math Inc. Teacher's Guide 4A Chapter 4 119

Alex thinks of 20 as 2 × 10. He multiplies 34 first by 2 and then by 10.

Demonstrate if needed:

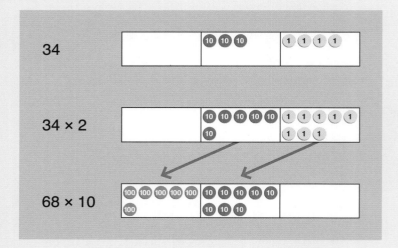

Mei thinks that since the number 20 is 2 tens, she can multiply 34 by 2. She relates that idea to the standard algorithm.

Because 20 is 2 tens, and any single-digit number of tens ends in one zero, a short cut is to multiply 34 by 2 and then append a zero.

Model the language:

"34 × 2 tens is 68 tens, so we can write a 0 in the ones column, and write 8 tens and 6 hundreds in the appropriate columns."

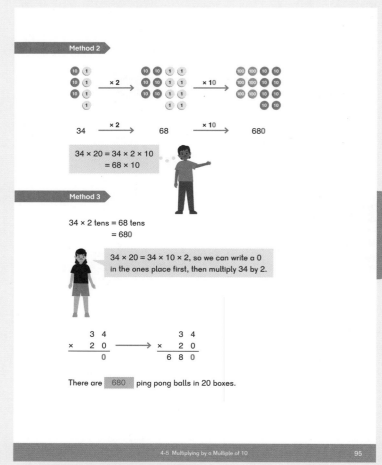

Do

① Students may notice they can just multiply 6 × 7 and append two zeros to the product. Explain to them why this works:

60 × 70 = 6 × 10 × 7 × 10

Since we can multiply in any order, we can change the order of the numbers in the equation:

6 × 7 × 10 × 10

Next, multiply the numbers together to make an easier problem:

42 × 100 = 4,200

④ These problems build on the idea that anytime we multiply a whole number by a whole number with a 0 in the ones place, the product will also have a 0 in the ones place.

We can write a 0 in the ones place in the product and then multiply by the digit in the tens place.

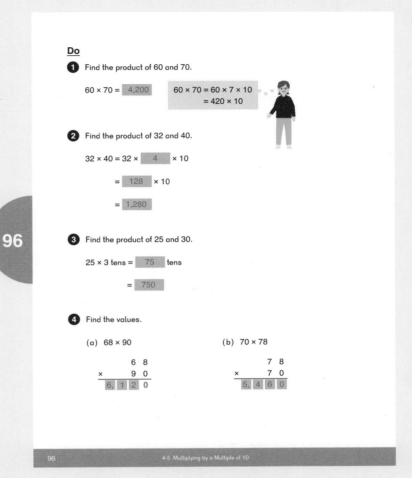

⑤–⑦ These problems follow the same sequence as ❶–❹, however now with three-digit numbers.

⑤ Students may notice you can just multiply 6 × 7 and append three zeros.

600 × 70 = 6 × 100 × 7 × 10

or

6 × 7 × 100 × 10 = 42 × 1,000 = 42,000

⑦ Sofia points out that the regrouping marks can go above the digit in the place multiplied next so that students do not forget to add that amount.

While 6 × 3 tens is actually 180, we write the regrouped 100 as a 1 above the tens place to remind us to add it to the product of 8 tens and 3 tens.

⑧ Students may use any method to find the products. The emphasis is on understanding multiplying by multiples of 10. Ask students: "Are you finding a number of hundreds? Tens? Thousands?"

Have students share their solutions for some of the problems.

Activity

▲ Closest to 3,000

Materials: Number Cards (BLM) 0–9, four sets, or deck of cards with face cards removed

Place the deck of cards facedown in the middle of the players. In each round, a player draws three cards and creates a two-digit number with two of the cards. The remaining card represents the number of tens. The player multiplies the two-digit number by the number of tens.

Example:

A player draws 3, 4, and 7, and creates the number 73. She multiplies 73 by 4 tens.

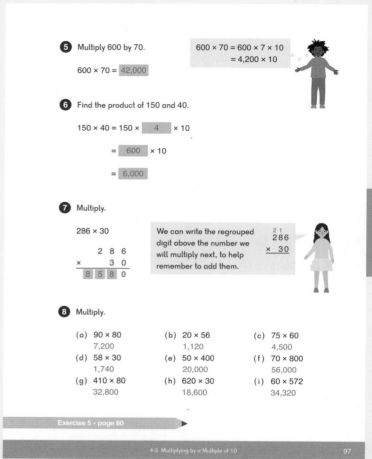

When all players have completed their equations, they compare their products. The player with a product closest to 3,000 wins a point.

The first player to get 10 points is the winner.

Exercise 5 • page 80

Lesson 6 Multiplying by a 2-Digit Number — Part 1

Objective

- Multiply a two-digit number by a two-digit number using the standard algorithm.

Lesson Materials

- Place-value discs

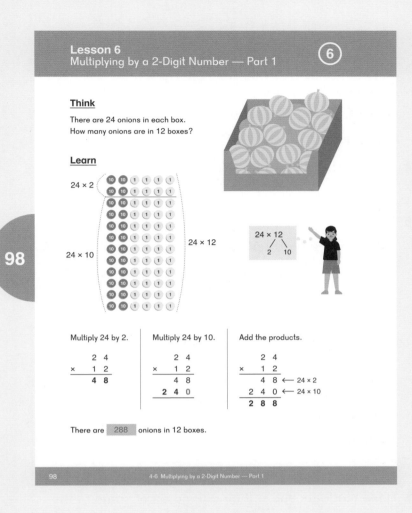

Think

Pose the **Think** problem. Students may solve with place-value discs, a picture, or calculations.

Ask students:

- "How is this problem different from the ones you solved in the previous lesson?" (We are not multiplying by a multiple of ten.)
- "How can you use what you already know to solve the problem?"

Have students estimate the product before computing. Discuss student strategies for solving the problem.

Learn

Work through the **Think** problem with students as demonstrated in **Learn**.

To understand the algorithm, Mei splits 12 into 2 and 10 and multiplies 24 by both of these numbers. The two partial products are then added together.

When numbers are written in the vertical format, the product of 24 × 2 is recorded first, then the product of 24 × 10 is recorded below. The two partial products are then added together.

Do

① The algorithm breaks a multi-digit calculation into single-digit calculations. Instead of multiplying 32 × 20, Sofia can just write a 0 in the ones place and multiply 32 × 2.

② Dion reminds students where in the algorithm to record the regrouping of 5 × 7.

③ Emma helps students stay organized by showing how to record the regrouping of the product of 8 × 4 tens.

While 8 × 4 tens is actually 320, we write the regrouped 300 as a 3 above the tens place to remind us to add it to the product of 5 tens × 4 tens.

Students who need additional help determining which regrouped digit to add can cross off the regrouped 7 tens from 8 × 9 before writing the 3 from 8 × 4 tens (32 tens).

⑤ Have students share their estimates of some of the problems.

Activity

▲ Greatest Product
Modified from Lesson 3

Materials: Number Cards (BLM) 0–9, four sets, or deck of cards with face cards removed

Place the deck of cards facedown in the middle of the players. On each turn, a player draws four cards and creates 2 two-digit numbers using the digits from the four cards. Players multiply the 2 two-digit numbers together.

When each player has had a turn, players compare their products. The player with the greatest product gets a point. The first player to get 10 points is the winner.

Exercise 6 · page 82

Lesson 7 Multiplying by a 2-Digit Number — Part 2

Objective

- Multiply a three-digit number by a two-digit number.

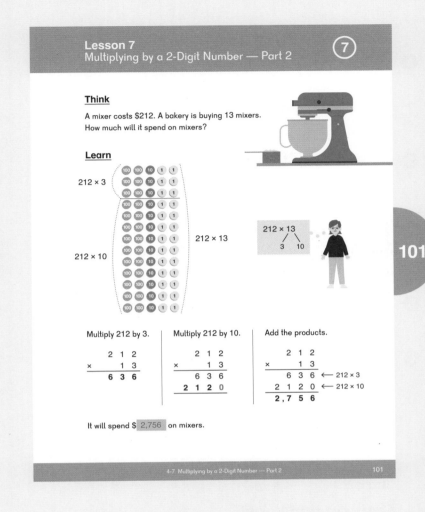

Think

Pose the **Think** problem. Have students estimate the product first.

Ask students:

- "How is this problem different from the ones you solved in the previous lesson?" (There are three digits that will be multiplied.)
- "How can you use what you already know to solve the problem?"

Discuss student strategies for solving the problem.

Learn

Work through the **Think** problem with students as demonstrated in **Learn**.

To understand the algorithm, Emma splits 13 into 3 and 10 and multiplies 212 by both of these numbers. The two partial products are then added together.

When numbers are written in the vertical format, the product of 212 × 3 is recorded first, then the product of 212 × 10 is recorded below.

Using an area model may help students connect the partial products to the algorithm. These models, however, are generally more work than the algorithm.

Do

1. The algorithm breaks a multi-digit calculation into single-digit calculations. Instead of multiplying 30 × 2, Alex can just write a 0 in the ones place and multiply 3 × 2.

2. Sofia reminds students to add the regrouped ten.

3. If students have difficulty remembering the regrouped digits, they can write them on top.

   Dion helps students stay organized by showing how to record the regrouping of the product of 9 × 5 tens.

5. Have students share their solutions for some of the problems.

6. A bar model with 28 boxes, each with a value of $155, would be cumbersome to draw. Students should know that this is a multiplication problem and use the algorithm to solve.

Activity

▲ **Greatest and Least**

Materials: Number Cards (BLM) 1–9

Choose five of the digits 1–9 once to find the greatest possible product.

Do the same to find the least possible product.

Exercise 7 • page 86

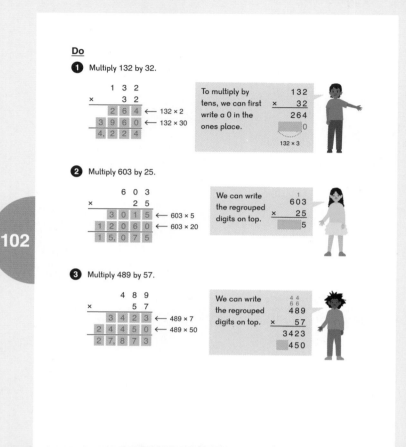

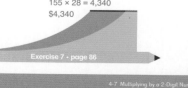

126   Teacher's Guide 4A Chapter 4   © 2019 Singapore Math Inc.

Lesson 8 Practice B

Objective

- Practice multiplication.

2 Have students share their estimates for solving some of the problems and state whether the product will be greater or less than their estimates, and why.

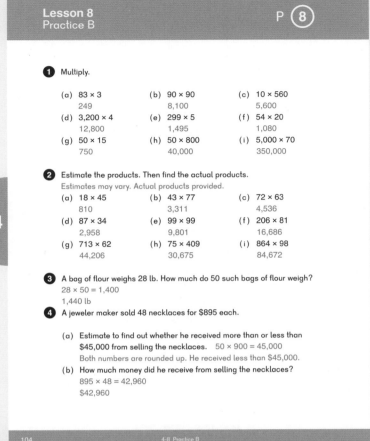

Lesson 8 Practice B — P 8

1 Multiply.
- (a) 83 × 3   249
- (b) 90 × 90   8,100
- (c) 10 × 560   5,600
- (d) 3,200 × 4   12,800
- (e) 299 × 5   1,495
- (f) 54 × 20   1,080
- (g) 50 × 15   750
- (h) 50 × 800   40,000
- (i) 5,000 × 70   350,000

2 Estimate the products. Then find the actual products.
Estimates may vary. Actual products provided.
- (a) 18 × 45   810
- (b) 43 × 77   3,311
- (c) 72 × 63   4,536
- (d) 87 × 34   2,958
- (e) 99 × 99   9,801
- (f) 206 × 81   16,686
- (g) 713 × 62   44,206
- (h) 75 × 409   30,675
- (i) 864 × 98   84,672

3 A bag of flour weighs 28 lb. How much do 50 such bags of flour weigh?
28 × 50 = 1,400
1,440 lb

4 A jeweler maker sold 48 necklaces for $895 each.
- (a) Estimate to find out whether he received more than or less than $45,000 from selling the necklaces.   50 × 900 = 45,000
Both numbers are rounded up. He received less than $45,000.
- (b) How much money did he receive from selling the necklaces?
895 × 48 = 42,960
$42,960

❺ Students should use mental math to find the total cost of each group of items.

❻ Students may estimate:
- 400 × 12 = 4,800
  22,300 + 4,800 = 27,100
- 350 × 12 = 3,500 + 700 = 4,200
  22,300 + 4,200 = 26,500

❼ Discuss the bar model with students.

Ask students:
- "What information is given in the problem and how is it represented on the model?"
- "Why are there dotted lines dividing the boxes that represent the chickens?"
- "How can we find the value of 1 unit?"

Encourage students to read the questions carefully so that they realize they do not need to find the number of rabbits or chickens.

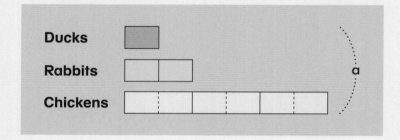

(a) If 1 unit represents 1,682 animals, and there are 9 total units:

1 unit ⟶ 1,682
9 units ⟶ 1,682 × 9 = 15,138
There are 15,138 animals in all.

(b) Students can count the units to find the difference between the number of birds and the number of rabbits.

There are 7 units of birds and 2 units of rabbits.

7 units − 2 units = 5 units
5 units ⟶ 1,682 × 5 = 8,410
There are 8,410 more chickens and ducks than rabbits.

---

❺ The students in the accounting club are helping the gym teacher buy basketball equipment by calculating the costs.

Item	Number Needed	Cost for 1
Backboard	4	$550
Backboard adjustment system	4	$1,200
Basketball	15	$34

How much will it cost to buy all of the equipment?
550 × 4 = 2,200
2,200 + 4,800 + 510 = 7,510
$7,510
1,200 × 4 = 4,800
34 × 15 = 510

❻ A sailboat costs $22,299 to buy and $358 a month to dock at the marina. Estimate to find out if $25,000 is enough to buy and dock the boat for the first year.
Estimate low: 300 × 12 = 3,600; 22,000 + 3,600 = 25,600
No.

❼ There are 1,682 ducks on a farm. There are twice as many rabbits as ducks. There are three times as many chickens as rabbits.

(a) How many ducks, rabbits, and chickens are there altogether?
1,682 × 9 = 15,138; 15,138 altogether
(b) How many more chickens and ducks altogether are there than rabbits?
1,682 × 5 = 8,410
8,410 more chickens and ducks than rabbits

⑨ Have students draw the model as shown in the answer overlay to see that Jamal earns $780 more each month than he did last year. Then have students find how much more money Jamal earns in total this year by multiplying 780 × 12.

⑩ If the quantity of nails and screws decreases by 510 and 75, respectively, the difference in decrease is equal to 1 Before unit: 510 − 75 = 435.

There were 2 units of nails at first:

1 unit ⟶ 435
2 units ⟶ 435 × 2 = 870
There were 870 nails in the bin at first.

Exercise 8 • page 90

Brain Works

★ **Mobiles**

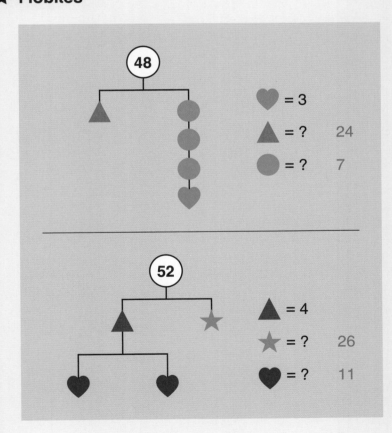

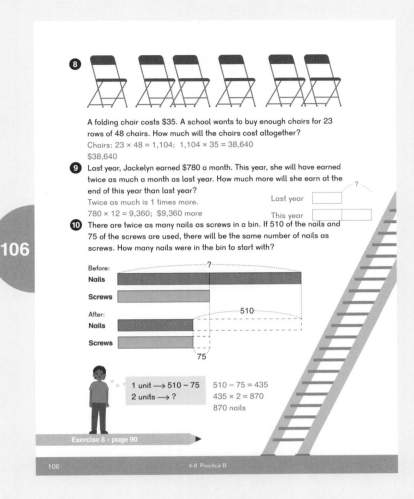

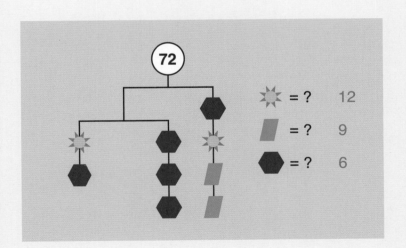

Exercise 1 • pages 67–68

Chapter 4 Multiplication

Exercise 1

Basics

1. (a) 12 × 3 = 30 + 6 = [36]
      (10, 2)
   (b) 12 hundreds × 3 = [36] hundreds = [3,600]
   (c) 12 thousands × 3 = [36] thousands = [36,000]

2. (a) 35 × 4 = [120] + 20 = [140]
      (30, 5)
   (b) 350 × 4 = [1,400]
   (c) 3,500 × 4 = [14,000]
   (d) 35,000 × 4 = [140,000]

3. (a) 100 × 7 = [700]   (b) 200 × 7 = [1,400]
       99 × 7 = [693]   199 × 7 = [1,393]
   (c) 300 × 7 = [2,100]   (d) 3,000 × 7 = [21,000]
       299 × 7 = [2,093]   2,999 × 7 = [20,993]

Practice

4. Use mental calculation to find the products.
   (a) 100 × 6 = [600]   (b) 150 × 9 = [1,350]
   (c) 40,000 × 8 = [320,000]   (d) 1,300 × 5 = [6,500]
   (e) 2,500 × 4 = [10,000]   (f) 910 × 2 = [1,820]
   (g) 15,000 × 3 = [45,000]   (h) 7,200 × 8 = [57,600]
   (i) 5 × 42,000 = [210,000]   (j) 7 × 3,300 = [23,100]
   (k) 98 × 6 = [588]   (l) 5 × 399 = [1,995]
   (m) 3 × 599 = [1,797]   (n) 6,999 × 4 = [27,996]

5. A 24-foot-wide frame for a greenhouse costs $4,999. A farm wants to buy 5 of them. What will be the total cost?
   5 × 4,999 = 24,995
   $24,995

6. The pediatric clinic needs 250 bandages a week. The central clinic needs 600 bandages a week. How many bandages should the two clinics order to have enough for 8 weeks?
   8 × 250 = 2,000   or   250 + 600 = 850
   8 × 600 = 4,800   8 × 850 = 6,800
   2,000 + 4,800 = 6,800

   6,800 band-aids

Exercise 2 • pages 69–72

Exercise 2

Basics

1 (a) Complete the following estimates for the product of 5,172 and 4.

5,172 × 4	5,172 × 4
↓	↓
5,000 × 4 = 20,000	5,200 × 4 = 20,800

(b) Fill in the missing numbers or digits for each calculation method.

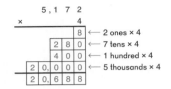

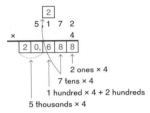

(c) Compare the estimates to the actual product.
Which estimate was lower? 20,000
Which estimate was higher? 20,800
Which estimate was closer? 20,800

Practice

2 (a) Dion estimated the product of 3,521 and 3 to be 12,000. With what number did he replace 3,521? 4,000

(b) Mei estimated the product of 3,521 and 3 to be 9,900. With what number did she replace 3,521? 3,300

(c) Whose estimate will be closer to the actual product?
Mei's estimate.

(d) Find the product of 3,521 and 3.

	3,	5	2	1
×				3
1	0,	5	6	3

3 Is 9,183 × 3 greater than or less than 25,000?
Estimate: 9,000 × 3 = 27,000
Greater than 25,000

4 The product of 4,962 and 8 is closest to which multiple of 10,000?
5,000 × 8 = 40,000
40,000

5 Are the following products reasonable? Why or why not?

(a) 7 × 986 = 6,902
7 × 1,000 = 7,000
Yes, it is close to the estimate.

(b) 71,081 × 8 = 57,448
70,000 × 8 = 560,000
No. The product should have 6 digits.

6 Estimate and then find the exact product. Estimates may vary.

(a) 8,292 × 3 ≈
8,292 × 3 = 24,876

	8,	2	9	2
×				3
2	4,	8	7	6

(b) 6,710 × 8 ≈
6,710 × 8 = 53,680

	6,	7	1	0
×				8
5	3,	6	8	0

(c) 6,114 × 7 ≈
6,114 × 7 = 42,798

	6,	1	1	4
×				7
4	2,	7	9	8

(d) 11,771 × 6 ≈
11,771 × 6 = 70,626

	1	1,	7	7	1
×					6
	7	0,	6	2	6

7 A backhoe costs $12,199. An excavator costs $165,599. A construction company wants to buy 4 backhoes and 1 excavator. What is the total cost?
4 × 12,199 = 48,796
165,599 + 48,796 = 214,395
$214,395

Challenge

8 Find the products. Do you notice a pattern?

1,089 × 1 = 1,089
1,089 × 2 = 2,178
1,089 × 3 = 3,267
1,089 × 4 = 4,356
1,089 × 5 = 5,445
1,089 × 6 = 6,534
1,089 × 7 = 7,623
1,089 × 8 = 8,712
1,089 × 9 = 9,801

Pattern: Each number is 1,000 and 100 more, and 10 and 1 less, than the previous number.

9 In the following problem, the letters S, T, O, and P stand for different digits. What is the number POTS?

```
    S T O P          3 3
  ×       4       2, 1 7 8
  ---------     ×       4
    P O T S      ---------
                  8, 7 1 2
```
POTS = 8,712

Since the answer is 4-digits, S × 4 < 10. S could be 1 or 2. Since P × 4 = S, S is even. So S = 2 and P = 8. There is no regrouping from the hundreds place so T × 4 < 10. T = 1. Since 8 × 4 = 32, the ones digit of O × 4 + 3 is 1. Possible values for O are 0, 3, 5, 6, 7, or 9. Only 7 works.

Exercise 3 • pages 73–76

Exercise 3

Basics

1 (a) Complete the following estimations for the product of 2,745 and 8.

2,745 × 8 → 3,000 × 8 = 24,000

2,745 × 8 → 2,500 × 8 = 20,000

(b) Fill in the missing numbers or digits for each calculation method.

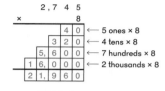

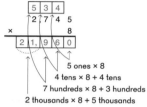

5 ones × 8
4 tens × 8 + 4 tens
7 hundreds × 8 + 3 hundreds
2 thousands × 8 + 5 thousands

(c) Compare the estimates to the actual product.
Which estimate was lower? 20,000
Which estimate was higher? 24,000
Which estimate was closer? 20,000

Practice

2 (a) Alex estimated the product of 52,891 and 4 to be 200,000. With what number did he replace 52,891?
50,000

(b) Sofia estimated the product of 52,891 and 4 to be 212,000. With what number did she replace 52,891?
53,000

(c) Whose estimate will be closer to the actual product?
Sofia's estimate

(d) Find the product of 52,891 and 4.

	5	2	8	9	1
×					4
	2	1	1	5	6	4

3 Is the product of 84,984 and 5 closer to 400,000 or 500,000?
85,000 × 5 = 425,000
400,000

4 Circle the number that is equal to 8,563 × 6 without calculating the exact product.

524,318   **51,378**   51,372   5,372

5 Which of the following gives the greatest product?

6,953 × 6   5,235 × 8   **5,673 × 9**

6 Estimate and then find the exact product. *Estimates may vary.*

(a) 7,884 × 4 ≈
7,884 × 4 = 31,536
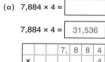

(b) 3,482 × 5 ≈
3,482 × 5 = 17,410

(c) 6,908 × 7 ≈
6,908 × 7 = 48,356

(d) 27,448 × 6 ≈
27,448 × 6 = 164,688

7 A leap year occurs every 4 years. A standard year has 365 days and a leap year has 366 days. How many days are in 4 consecutive years?

365 × 3 = 1,095   or   365 × 4 = 1,460
1,095 + 366 = 1,461   1,460 + 1 = 1,461
1,461 days

Challenge

8 Find the products. Do you notice a pattern?

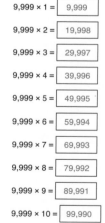

Pattern: Each number is 10,000 more and 1 less than the previous number.

9 In the following problem, the letters G, R, E, A, and T stand for different digits. What is the number GREAT?

```
    1 G R E A T              1   2 1 2
  ×           3            1 4 2, 8 5 7
  -----------                      × 3
    G R E A T 1   GREAT = 42,857   ---------
                                4 2 8, 5 7 1
```

Since 1 is in the ones place of the product, T = 7. The ones digit of A × 3 + 2 = 7, so A = 5. Similarly, E = 8, R = 2, and T = 4.

Teacher's Guide 4A Chapter 4   © 2019 Singapore Math Inc.

Exercise 4 • pages 77–79

Exercise 4

Check

1 Is the product of 87,984 and 6 closer to 500,000 or 600,000?
90,000 × 6 = 540,000
500,000

2 Estimate to arrange the expressions in order from least to greatest.

16,953 × 3	21,992 × 2	7,673 × 7	4,813 × 9
A	B	C	D

D, B, A, C

3 Estimate and then find the exact product.   Estimates may vary.

(a) 3,806 × 7 ≈
3,806 × 7 = 26,642

(b) 9,458 × 8 ≈
9,458 × 8 = 75,664

(c) 73,987 × 2 ≈
73,987 × 2 = 147,974

4 There are 5,280 feet in 1 mile. How many feet are in 8 miles?
5,280 × 8 = 42,240
42,240 feet

5 An artist sold 5 paintings for $3,410 each. If the frame, canvas, paint, and other material for each painting cost $155, how much profit did he make from the paintings after subtracting the cost of materials?
3,410 − 155 = 3,255
3,255 × 5 = 16,275
$16,275

6 An aquarium had 3,876 visitors one day. Each visitor was given 2 drink coupons and 3 food coupons to use at the food court. How many coupons were given out?
Coupons given to each visitor: 2 + 3 = 5
3,876 × 5 = 19,380
19,380 coupons

7 To raise money to rescue marine animals, 3 corporations pledged to give $55,600 each to the charity if their total contributions were matched by other donors. The amount was matched and an additional $4,590 was raised. How much money was raised for the charity in all?
55,600 × 6 = 333,600
333,600 + 4,590 = 338,190
$338,190

Challenge

8 Find the missing digits.

(a)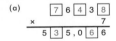
   7 6 4 3 8
  ×   7
   5 3 5, 0 6 6

(b)
   9 1, 4 0 6
  ×   4
   3 6 5, 6 2 4

9 Alex's estimate: 6 × 896 ≈ 10 × 896 = 8,960
Emma's estimate: 6 × 896 ≈ 6 × 900 = 5,400

Both of them increased one factor by 4, but one of the estimates is much closer to the actual answer. Without calculating the answer, determine which estimate is closer and why.
Emma's estimate is closer. An increase of 6 by 4 has a greater impact on the product than an increase of 896 by 4.

10 In a large tank at an aquarium, there were twice as many jellyfish as water snails. After 430 snails were removed from the tank, there were 3 times as many jellyfish as water snails. How many jellyfish were there?
430 × 6 = 2,580
2,580 jellyfish

Check:
2,580 ÷ 3 = 860
860 + 430 = 1,290
1,290 × 2 = 2,580

Before
Jellyfish
Water Snails

After
Jellyfish
Water Snails
430

Exercise 5 • pages 80–81

Exercise 5

Basics

1. (a) 90 × 10 = 900 (b) 900 × 10 = 9,000
   (c) 46 × 10 = 460 (d) 46 × 100 = 4,600
   (e) 80 × 10 = 800 (f) 800 × 100 = 80,000

2. (a) 6 × 3 = 18 (b) 60 × 30 = 1,800
   (c) 600 × 3 = 1,800 (d) 6 × 300 = 1,800
   (e) 600 × 30 = 18,000 (f) 60 × 300 = 18,000

3. (a) 62 × 30 = 62 × 3 × 10
       = 186 × 10
       = 1,860
   (b) 620 × 30 = 620 × 3 × 10
       = 1,860 × 10
       = 18,600

4. Write the missing digits.

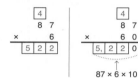
87 × 6 × 10

870 × 6 × 10

Practice

5. Multiply.

   70 × 4 = 280
   70 × 40 = 2,800
   87 × 5 = 435
   87 × 50 = 4,350

   674 × 6 = 4,044
   674 × 60 = 40,440
   395 × 9 = 3,555
   395 × 90 = 35,550

   950 × 8 = 7,600
   950 × 80 = 76,000
   984 × 7 = 6,888
   984 × 70 = 68,880

6. A snail has 33 teeth in 80 rows. How many teeth does the snail have?
   33 × 80 = 2,640
   2,640 teeth

7. How many minutes are in 28 hours?
   28 × 60 = 1,680
   1,680 minutes

8. How many seconds are in 40 hours?
   40 × 60 = 2,400
   2,400 × 60 = 144,000
   144,000 seconds

Exercise 6 • pages 82–85

Exercise 6

Basics

1 (a) Estimate the product of 64 and 57.   Estimates may vary.

64 × 57

□ × □ = □

(b) Write the missing numbers or digits.

```
        2
        2
      6 4
  ×   5 7
      4 4 8  ← 64 × 7
    3,2 0 0  ← 64 × 50
    3,6 4 8
```

2 (a) Alex estimated the product of 95 and 75 to be 8,000. With what number could he have replaced each factor?
He replaced 95 with 100 and 75 with 80.
(b) Mei estimated the product of 95 and 75 to be 7,200. With what number could she have replaced each factor?
She replaced 95 with 90 and 75 with 80.
(c) Whose estimate will be closer to the product?
Mei's estimate
(d) Find the product of 95 and 75.

```
      9 5
  ×   7 5
      4 7 5
    6,6 5 0
    7,1 2 5
```

Practice

3 Circle the number that is equal to 47 × 57 without calculating the exact product.

| 1,799 | 3,829 | 2,679 | 279 |

4 Estimate and then find the exact product.   Estimates may vary.

(a) 27 × 85 ≈ □
27 × 85 = 2,295

```
      2 7
  ×   8 5
      1 3 5
    2,1 6 0
    2,2 9 5
```

(b) 48 × 48 ≈ □
48 × 48 = 2,304

```
      4 8
  ×   4 8
      3 8 4
    1,9 2 0
    2,3 0 4
```

(c) 65 × 72 ≈ □
65 × 72 = 4,680

```
      6 5
  ×   7 2
      1 3 0
    4,5 5 0
    4,6 8 0
```

(d) 91 × 38 ≈ □
91 × 38 = 3,458

```
      9 1
  ×   3 8
      7 2 8
    2,7 3 0
    3,4 5 8
```

5 24 cookies can be baked on one baking sheet. The cafeteria cook bakes 25 sheets of cookies. Is this enough for 500 people to have at least one cookie?
Estimate is sufficient: 20 × 25 = 500
Yes. There will be more than 500.

6 Tickets to the aquarium cost $45 per adult and $35 per child. There are 9 adults and 64 children in a group.

(a) Without calculating the actual cost, determine whether the tickets for the entire group will cost less than $3,000.
Estimate high:
50 × 10 = 500
40 × 60 = 2,400
500 + 2,400 = 2,900
Yes.

(b) What is the total cost of the tickets for the group?
9 × 45 = 405
64 × 35 = 2,240
405 + 2,240 = 2,645
$2,645

7 Danios, barbs, and loaches are types of fish that live in Asian rivers. There are 63 danios in an Asian river fish habitat at the aquarium. There are five times as many barbs as danios and three times as many loaches as barbs. How many of these three types of fish are in the habitat?
1 unit → 63
21 units → 63 × 21 = 1,323
1,323 fish

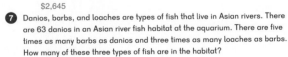

8 What will be the ones digit in the product of each of the following expressions? Is the product even or odd?

	Ones Digit	Even or Odd
2 × 3	6	Even
42 × 33	6	Even
5 × 3 × 3	5	Odd
85 × 33 × 3	5	Odd

Students should realize that they only have to multiply ones digits to find the ones digit of the products.

Challenge

9 Find the missing digits.

(a)
```
      3 4
  ×   1 8
      2 7 2
      3 4 0
      6 1 2
```

(b)
```
      1 7
  ×   6 5
        8 5
    1,0 2 0
    1,1 0 5
```

10 Tank A had 3 times as many fish as Tank B. After 240 fish were moved from Tank A to Tank B, Tank A had the same number of fish as Tank B. How many fish are in both tanks altogether?
240 × 4 = 960
960 fish

Before
A □ □
B □

After
A □ 240
B 240

© 2019 Singapore Math Inc.   Teacher's Guide 4A Chapter 4

Exercise 7 • pages 86–89

Exercise 7

Basics

1 (a) Estimate the product of 537 and 28.   Estimates may vary.

537 × 28
↓   ↓
[ ] × [ ] = [ ]

(b) Fill in the missing numbers or digits.

```
        1
      2 5
      5 3 7
  ×     2 8
    4, 2 9 6  ← 537 × 8
  1 0, 7 4 0  ← 537 × 20
  1 5, 0 3 6
```

2 (a) Sofia estimated the product of 649 and 15 to be 12,000. With what numbers could she have replaced each factor?
600 and 20
(b) Emma estimated the product of 649 and 15 to be 13,000. With what numbers could she have replaced each factor?
650 and 20
(c) Whose estimate will be closer to the actual product?
Sofia's estimate
(d) Find the product of 649 and 15.

```
      6 4 9
  ×     1 5
    3, 2 4 5
    6, 4 9 0
    9, 7 3 5
```

Practice

3 Circle the number that is equal to 845 × 62 without calculating the exact product.

[56,390]   [5,230]   [52,390]   [25,390]

4 There are 12 inches in a foot and 5,280 feet in a mile. Which is longer, 50,000 inches or 1 mile?
12 × 5,280 = 63,360
There are 63,360 inches in a mile.
1 mile is longer.

5 Estimate and then find the exact product.   Estimates will vary.

(a) 638 × 48 ≈ [ ]

638 × 48 = 30,624

```
        6 3 8
  ×       4 8
    5, 1 0 4
  2 5, 5 2 0
  3 0, 6 2 4
```

(b) 512 × 72 ≈ [ ]

512 × 72 = 36,864

```
        5 1 2
  ×       7 2
    1, 0 2 4
  3 5, 8 4 0
  3 6, 8 6 4
```

(c) 821 × 58 ≈ [ ]

821 × 58 = 47,618

```
        8 2 1
  ×       5 8
    6, 5 6 8
  4 1, 0 5 0
  4 7, 6 1 8
```

(d) 763 × 65 ≈ [ ]

763 × 65 = 49,595

```
        7 6 3
  ×       6 5
    3, 8 1 5
  4 5, 7 8 0
  4 9, 5 9 5
```

6 Each box contains 275 bandages. There were 14 full boxes and one box that contained 123 bandages. 3 full boxes were used. How many bandages are left?
Full boxes left: 14 − 3 = 11
Bandages left:
11 × 275 = 3,025
3,025 + 123 = 3,148
3,148 bandages

7 (a) Is the product of 4 × 9 × 7 × 1 odd or even? What is the ones digit of the product?
4 × 9 × 7 × 1 = 252
If any of the factors is even, then the product will be even.
Even
Ones digit: 2

(b) Is the product of 4 × 419 × 17 × 11 odd or even? What is the ones digit of the product?
It is only necessary to multiply the ones digits to determine the ones digit of the product.
Even
Ones digit: 2

8 Write down any 3-digit number. Multiply it by 11. Multiply your answer by 91. What do you notice about your final answer? Try other numbers.
If the chosen number is abc, with a, b, and c representing any digit, then the final answer is abc,abc.
Students are not required to explain why this occurs.
11 × 91 = 1,001
1,001 × abc = (1,000 + 1) × abc = abc000 + abc

Challenge

9 Find the missing digits.

(a)
```
        1 4 6
  ×       7 1
        1 4 6
  1 0, 2 2 0
  1 0, 3 6 6
```

(b)
```
        8 7 3
  ×       9 3
    2, 6 1 9
  7 8, 5 7 0
  8 1, 1 8 9
```

10 Consider the following examples.

28 × 25 = 7 × 4 × 25 = 7 × 100 = 700

36 × 50 = 18 × 2 × 50 = 18 × 100 = 1,800

18 × 35 = 9 × 2 × 5 × 7 = 9 × 7 × 10 = 630

Use similar methods to find the following products. Look for factors that together have a product of 10, 100, or 1,000.

(a) 16 × 25 = [400]   (b) 25 × 44 = [1,100]

(c) 14 × 45 = [630]   (d) 26 × 15 = [390]

(e) 75 × 12 = [900]   (f) 50 × 168 = [8,400]

(g) 250 × 36 = [9,000]   (h) 125 × 16 = [2,000]

Exercise 8 • pages 90–94

Exercise 8

Check

1 Multiply and use the answers to complete the cross number puzzle on the next page.

Across

```
    862           739            89
×    45        ×   42         ×  42
 38,790         31,038         3,738
     A              D              F

  9,607          989            773
×     8        ×  89          ×  58
 76,856        88,021         44,834
     G              J              K
```

Down

```
  5,897           96             69
×     6        ×  81          ×  24
 35,382         7,776          1,656
     A              B              C

    438          874             48
×    70        ×  62          ×  58
 30,660        54,188          2,784
     E              H              I
```

Cross number puzzle completed with:
- A across: 38790 (A₃ 8 7 9 0)
- B across: 7 5 (B₇ 5)
- C down: C₁
- D across: D₃ 1 0 E₃ 8, 6
- F across: F₃ 7 3 8, 0, 5
- G across: 6, 2, G₇ 6 8 H₅ 6
- I down: I₂
- J across: J₈ 8 0 2 1, 7
- 8, 8
- K across: K₄ 4 8 3 4

2 Write > or < in each ◯. Use estimation.

(a) 8,268 × 4 ⟩ 32,000
(b) 16 × 84 ⟨ 20 × 90
(c) 271 × 12 ⟩ 270 × 10
(d) 3,294 × 156 ⟨ 3,300 × 160
(e) 6,198 × 4 ⟩ 781 × 24
(f) 326 × 21 ⟨ 96 × 74

3 Is it possible to multiply a 2-digit whole number by a 2-digit whole number and get a number that is greater than 4 digits? Explain your answer.
No. The greatest possible product is 99 × 99 = 9,801, which is 4 digits.

4 The pumps and sand filters at the aquarium clean 942 gallons of water a minute. How many gallons do they clean in 45 minutes?
942 × 45 = 42,390
42,390 gallons

5 One of the buildings at the aquarium is getting new energy-saving light bulbs. Each light fixture holds 4 bulbs. There are 118 light fixtures on each floor. There are 4 floors in the building. How many light bulbs are needed?
118 × 4 × 4 = 1,888
1,888 light bulbs

6 The aquarium is selling T-shirts at a reduced price of $5 each to celebrate Earth Day. T-shirts are packaged 32 to a box. They ordered 19 boxes of large, 24 boxes of medium, and 22 boxes of small. If they sold all the T-shirts, how much money would they receive?
19 + 24 + 22 = 65
65 × 32 = 2,080
2,080 × 5 = 10,400
$10,400

Challenge

Consecutive numbers are numbers that come one after the other. For example, 3, 4, and 5 are three consecutive whole numbers.

7 What is the ones digit of the product of any 10 consecutive whole numbers?
0
Students should realize that they only have to concern themselves with the digit in the ones place, so they can look at single digits numbers. 10 consecutive numbers will always include either 0 or a multiple of 10, so their product will have 0 in the ones place.

8 What is the ones digit of the product of any 5 consecutive whole numbers?
0
Students may use trial and error using numbers within 20. Every set of 5 consecutive numbers will either contain 0 or a multiple of 10, or will contain one number with a 5 in the ones digit and another number that is even, resulting in a multiple of 10 when those two are multiplied together.

9 What are the possible ones digits for the products of any two consecutive whole numbers?
0, 2, 6
We only need to be concerned with the ones digit of the two numbers.
0 × 1 = 0, 1 × 2 = 2, 2 × 3 = 6, 3 × 4 = 12, 4 × 5 = 20, 5 × 6 = 30, 6 × 7 = 42, 7 × 8 = 56, 8 × 9 = 72, 9 × 10 = 90

10 The product of 4 consecutive whole numbers is 3,024. What are the numbers?
6, 7, 8, 9
Students may use guess and check.
None of the numbers can be 10 or 5, or the product would end in 0. At least one of them must be less than 10, or the product would have at least 5 digits. Since they are consecutive, they all need to be less than 10, or one of them will be 10.

Teacher's Guide 4A Chapter 4

11. Study the pattern. How many dots will be in the 24th rectangular number?

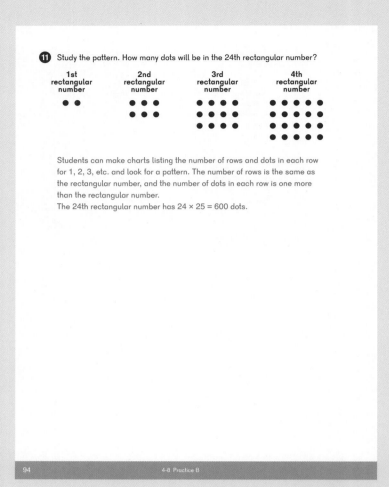

Students can make charts listing the number of rows and dots in each row for 1, 2, 3, etc. and look for a pattern. The number of rows is the same as the rectangular number, and the number of dots in each row is one more than the rectangular number.

The 24th rectangular number has 24 × 25 = 600 dots.

Chapter 5 Division　　　　　　　　　　　　　　　　　　　　Overview

Suggested number of class periods: 8–9

	Lesson	Page	Resources	Objectives
	Chapter Opener	p. 145	TB: p. 107	Investigate division.
1	Mental Math for Division	p. 146	TB: p. 108 WB: p. 95	Divide numbers up to three-digits by a one-digit number using a mental math method.
2	Estimation and Division	p. 148	TB: p. 111 WB: p. 97	Estimate quotients.
3	Dividing 4-Digit Numbers	p. 152	TB: p. 115 WB: p. 101	Divide up to a four-digit number by a one-digit number.
4	Practice A	p. 156	TB: p. 120 WB: p. 105	Practice division.
5	Word Problems	p. 158	TB: p. 122 WB: p. 109	Solve multi-step word problems involving multiplicative comparison.
6	Challenging Word Problems	p. 162	TB: p. 126 WB: p. 113	Extension: Solve multi-step word problems.
7	Practice B	p. 166	TB: p. 130 WB: p. 118	Practice division.
	Review 1	p. 168	TB: p. 132 WB: p. 121	Cumulative review of content from Chapters 1–5.
	Workbook Solutions	p. 170		

Chapter 5 Division

Notes

In Dimensions Math 3A students learned to divide numbers of up to three digits by one digit using the standard algorithm. They also learned some mental math strategies for simpler division problems.

In this chapter, students extend their understanding of the algorithm to four-digit numbers and apply their skills to a variety of word problems.

Mental Math for Division

In Lesson 1, students use place value concepts and calculation strategies from Chapter 1 Lesson 7 along with their knowledge of multiplication and multiples from Chapter 4 to solve division problems using mental math.

Multiples of tens, hundreds, and thousands can easily be solved by thinking of simple one-digit and two-digit computations. For example, 420 ÷ 3 can be thought of as 42 tens ÷ 3.

$$420 \div 3 \qquad 42 \div 3 = 10 + 4 = 14$$
$$\diagdown \diagup$$
$$30 \quad 12$$

42 tens ÷ 3 = 14 **tens** or 14**0**

Estimation and Division

Students will apply their knowledge of multiplication and multiples to estimate quotients in a division problem.

Estimation in division problems is often difficult because it requires more number sense than rounding to a given place. Students will use their knowledge of multiplication facts to find a convenient multiple of the divisor in order to estimate.

The goal is for students to find numbers close to those in the problem that they can calculate easily to find an estimate, and to use that estimate to check if their answers are reasonable. Estimation is an important skill for students to develop prior to Dimensions Math 5A, when they will be dividing by two-digit numbers.

Students should be able to explain whether their estimates will be greater or less than the actual quotient.

Standard Division Algorithm

The division algorithm requires an understanding of all four operations, as well as place value, regrouping, and estimation. It also looks different from the standard addition, subtraction, and multiplication algorithms. It begins with the digit in the highest place, rather than ones. The division algorithm involves dividing and regrouping the remainder for each place, one after the other.

Students will begin with a review of the algorithm with three-digit dividends, then move to four-digit dividends.

For a more basic review of the division algorithm, reference Dimensions Math 3A Chapter 6.

For reference, a general procedure for demonstrating the division algorithm with place-value discs is provided. Students should understand the reasoning for each step, including regrouping.

$$4{,}642 \div 3 \qquad 3\overline{)4{,}642}$$

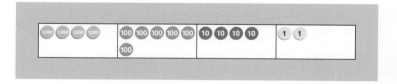

Divide 4 thousands by 3:

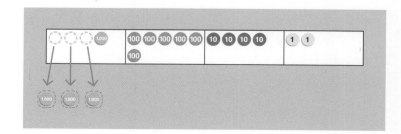

Chapter 5 Division

Notes

- 4 thousands divided into 3 equal groups is 1 thousand in each group with 1 thousand remaining. Write the digit 1 in the thousands place in the quotient.

- Of the 4 thousands, 3 have been divided (1 × 3 = 3). Write 3 below the 4 in the thousands place and find the difference, which is the thousand that still needs to be divided.

- Regroup the remaining 1 thousand as 10 hundreds. There are now 16 hundreds. Write a 6 in the hundreds place next to the remaining thousand.

Divide 16 hundreds by 3:

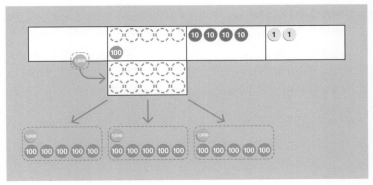

- 16 hundreds divided into 3 equal groups is 5 hundreds in each group with 1 hundred remaining. Write the digit 5 in the hundreds place in the quotient.

- Of the 16 hundreds, 15 were divided equally into groups (5 × 3 = 15). Write 15 below 16 and find the difference, which is the number of hundreds that still needs to be divided.

- Regroup the remaining 1 hundred as 10 tens. There are now 14 tens. Write 4 tens next to the remaining hundred.

Divide 14 tens by 3:

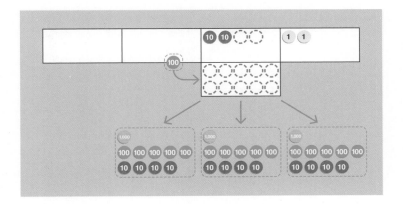

- 14 tens divided into 3 equal groups is 4 tens in each group with 2 tens remaining. Write the digit 4 in the tens place in the quotient.

- Of the 14 tens, 12 were divided equally into groups (4 × 3 = 12). Write 12 below the 14 tens, and subtract to find the difference, which is the tens that still need to be divided.

- Regroup the remaining 2 tens as 20 ones. There are now 22 ones.

Divide 22 ones by 3:

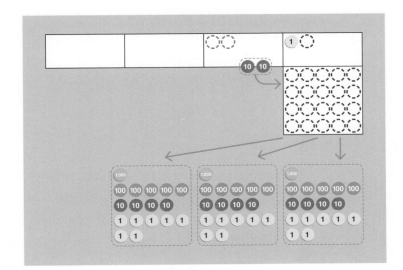

Chapter 5 Division

- 22 ones divided into 3 equal groups is 7 ones in each group with 1 one remaining. Write the digit 7 in the ones place in the quotient.
- Of the 22 ones, 21 were divided equally into groups (7 × 3 = 21). Write 21 below the 22 and subtract to find the difference, which is the ones that will not be divided further at this point.

```
        T H T O
        1 5 4 7
    3 ) 4 6 4 2
        3
        1 6
        1 5
          1 4
          1 2
            2 2
            2 1
              1
```

4,642 ÷ 3 is 1,547 R1. We can check the answer with multiplication:

3 × 1,547 + 1 = 4,642

An area model is a less obvious way to see the partial quotients and is only practical when there is no remainder. Its use is limited in division. Students find the missing factors from the area of smaller portions of the entire rectangle.

For example, 729 ÷ 3 = 243:

	600 ÷ 3 = 200	120 ÷ 3 = 40	9 ÷ 3 = 3
3	600	120	9

Bar Models

As with multiplication, it is expected that students have experience using bar models to help them solve word problems. If students are unfamiliar with bar models, reference Dimensions Math 3A Chapter 4 Lessons 7–9.

Next is a review of the two main types of models students learned in Dimensions Math 3: Part-whole Models and Comparison Models. There are two ways of showing a model in division problems, depending on whether the number of groups, or the quantity in each group, is given.

Part-whole Models

This type of bar model extends the understanding of multiplicative relationships to division. Each part is an equally sized group and is considered a "unit."

3 identical smart speakers cost $420.
How much does each smart speaker cost?

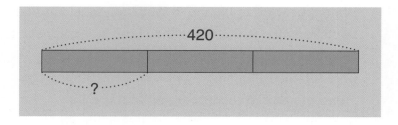

3 units ⟶ 420
1 unit ⟶ 420 ÷ 3 = 140
Each smart speaker costs $140.

Some problems will provide the quantity in each group, not the number of groups. While the division procedure is the same, the models differ.

Mei raised $408 for the art club by selling hats for $6 each. Estimate, and then find the number of hats she sold.

408 ÷ 6 = 68
Mei sold 68 hats.

Comparison Models

This type of bar model allows students to compare quantities. They are particularly useful in representing multi-step problems.

A full-size fridge costs $570. It costs 3 times as much as a compact fridge. How much does the compact fridge cost?

Chapter 5 Division

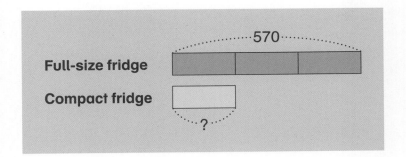

3 units ⟶ 570
1 unit ⟶ 570 ÷ 3 = 190
The compact fridge costs $190.

From this same model, students can also find:
How much do the two fridges cost altogether?

1 unit ⟶ 190
4 units ⟶ 4 × 190 = 760
The fridges cost $760 altogether.

When students are having difficulties solving a word problem, ask prompting questions:

- "What information do you know?"
- "What information do you need to find?"
- "How can you draw a model to show this information?"

In addition to making the numbers easier to calculate and removing numbers, there are other strategies that can help students understand word problems.

For example:

The sum of two numbers is 3,750. The second number is 180 less than twice the first number. What are the two numbers?

Present a simpler problem to students who are struggling:

The sum of two numbers is 30. The second number is 3 less than twice the first number. What are the two numbers?

Students should think, "If the second number was just 3 more, there would be 3 equal units."

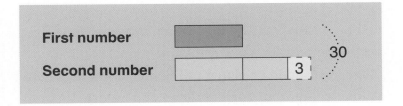

30 + 3 gives the value of 3 equal units.
3 units ⟶ 33
1 unit ⟶ 33 ÷ 3 = 11
The first number is 11.

2 units ⟶ 2 × 11 = 22
The second number is 22 − 3 = 19.

Once they can think about or draw the model, they can replace the simpler numbers with the ones in the original problem.

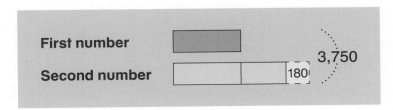

3,750 + 180 gives the value of 3 equal units.
3 units ⟶ 3,930
1 unit ⟶ 3,930 ÷ 3 = 1,310
The first number is 1,310.

2 units ⟶ 2 × 1,310 = 2,620
The second number is 2,620 − 180 = 2,440.

Challenging Word Problems

As problems become more complex, students may need to erase and redraw the bar models. Whiteboards allow students to quickly modify their drawings when first learning to draw bar models.

Students should read through each problem completely before drawing bar models on their own.

Chapter 5 Division

Notes & Materials

Learning to solve word problems is a vital skill that can be developed with discussion and practice. The bar model provides a way to start organizing the steps involved in solving a problem. Bar models help students see what computations can be used in the problem solving process.

Note: Lesson 6: Challenging Word Problems is an optional extension lesson that is included in order to allow students the opportunity to solve rich, higher level questions. All students should attempt to solve these problems. Encourage students to use the strategies taught throughout this chapter. If necessary, suggest easier numbers for students to use. Teacher discretion allows the opportunity for differentiation.

Materials

- Dry erase markers
- Place-value discs
- Play money
- Whiteboards

Blackline Masters

- Divvy Up Game
- Number Cards

Activities

Activities included in this chapter provide practice with division. They can be used after students complete the **Do** questions, or anytime additional practice is needed.

Chapter Opener

Objective

- Investigate division.

This **Chapter Opener** is a review of division covered in Dimensions Math 3A. Have students discuss mental math strategies that might help them solve the division questions.

This lesson may continue straight to Lesson 1.

Lesson 1 Mental Math for Division

Objective

- Divide numbers up to three-digits by a one-digit number using a mental math method.

Lesson Materials

- Play money or place-value discs

Think

Pose the **Think** problem and ask students to use a mental math method to find the cost of each smart speaker. If students cannot think of a method, provide them with play money or place-value discs and have them solve the problem concretely before trying mental math.

Discuss student methods.

Learn

Discuss the methods shown in **Learn** and have students compare their own methods with the methods shown in the textbook. Emma thinks of 420 as 42 tens and splits it into 30 tens and 12 tens. Both numbers are easily divisible by 3.

If students think of tens as the unit, they can just divide 42 by 3. Since answers will be in the tens, they need to append a 0 (14 × 10 = 140).

Alex wonders if there are other ways to divide 420 easily.

Examples of possible solutions:

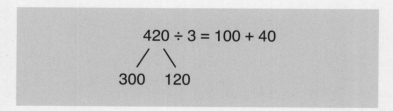

or

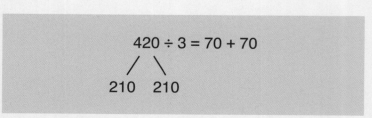

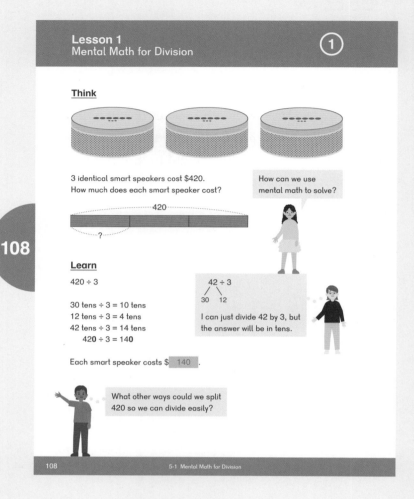

Because 6 = 3 × 2, another mental strategy that could be used is:

42 tens ÷ 6 = 7 tens
42 tens ÷ 3 = 7 tens × 2

146 Teacher's Guide 4A Chapter 5 © 2019 Singapore Math Inc.

Do

1 – 2 Students can use Dion's method combined with their knowledge of place value to easily see the patterns and relationships. 56 ÷ 8 is the same computation whether the number being divided by 8 is 56 ones, 56 tens, 56 hundreds, or 56 thousands.

56 **tens** ÷ 8 = 7 **tens**
56 **hundreds** ÷ 8 = 7 **hundreds**
56 **thousands** ÷ 8 = 7 **thousands**

Students may also see that once they find 56 ÷ 8 = 7, they can solve related problems such as 560 ÷ 8 = 70 by knowing that 560 is 56 tens. The quotient has the same number of zeros appended as the dividend compared to the numbers used in the computation:

56**0** ÷ 8 = 7**0**
5,6**00** ÷ 8 = 7**00**
56,**000** ÷ 8 = 7,**000**

3 In this problem, students cannot merely think of the number 800 without the zeros.

Sofia suggests thinking of 800 as 80 and splitting 80 into 50 and 30.

4 Some students may think of 96,000 as 66,000 and 30,000 because 66 = 11 × 6 and the 11 facts are interesting for students.

To solve:

$$96{,}000 \div 6 = 11{,}000 + 5{,}000$$
$$\underset{66\text{ thousand} \quad 30\text{ thousand}}{\diagup \quad \diagdown}$$

Exercise 1 • page 95

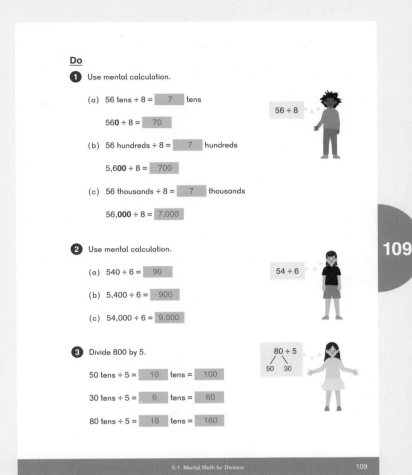

Lesson 2 Estimation and Division

Objective
- Estimate quotients.

Lesson Materials
- Place-value discs

Think

Provide students with place-value discs and discuss the **Think** problems. Have students estimate a number near 143 that is divisible by 3 and then have them use the discs to solve the problem. Discuss Emma's question.

Students may easily recall the long division algorithm. Allow students who need additional help to use the place-value discs.

Discuss student estimates and ideas for finding the actual amount.

Learn

Discuss the methods shown in **Learn** and have students compare their own methods with the methods shown in the textbook.

Dion points out that while the closest multiples of 3 tens to 143 are 120 and 150, 150 gives a closer estimate as it is closer to 143. Students should find a number close to the dividend that is easy to divide mentally and then think about whether it will be an overestimate or an underestimate.

This estimation work is important to help students later with the algorithm in the future.

While **Think** (b) does not specifically ask for a mental strategy, students may have found the actual amount using one.

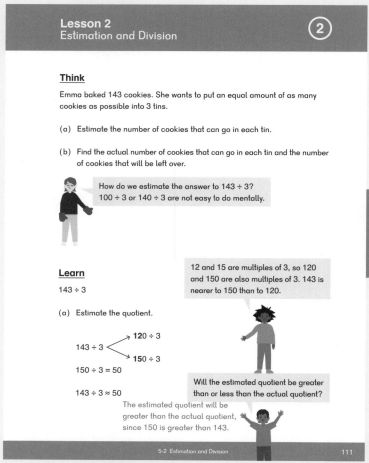

For example:

143 ÷ 3 is 40 + 7 remainder 2
      / | \
    120 21 2

This strategy works well until remainders are difficult to compute mentally.

Work through the **Think** problem as demonstrated in **Learn**.

Have students work along with place-value discs as the steps are modeled.

Discuss the regrouping with the discs and show students the written steps.

Students can check the answer with multiplication: 47 × 3 + 2 = 143.

Mei introduces a shorthand way of recording a remainder using a capital R.

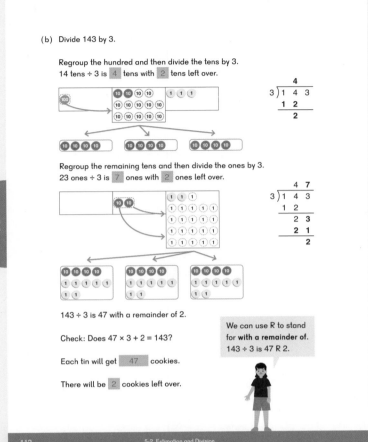

Do

1 (a) Discuss Dion's comments on the closest multiple of 300. He uses 900 to estimate his answer since 823 is closer to 900 than 600.

(b) Students who need additional practice should solve the problem with place-value discs first, then relate the steps they took with the discs to the division problem in the textbook.

Emma asks students to think about the number of digits in the quotient. Students should see that as 800 is more than 300, there will be a digit in the hundreds column, and so the quotient is a three-digit number.

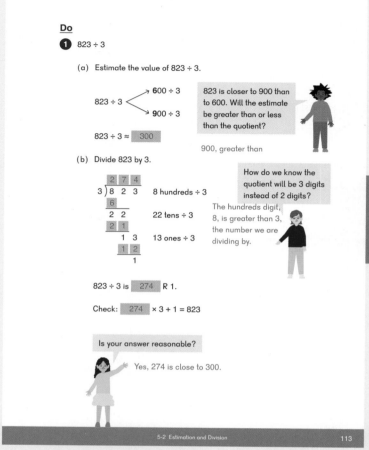

❷ Ask students why Dion chose 540 and 630 for his estimates. 568 is approximately 57 tens, and 57 is in between 2 multiples of 9: 54 and 63.

Students should see that since 568 is less than 900, there will not be a digit in the hundreds column. The hundreds will be regrouped as 56 tens. The answer will be a two-digit number.

❸ Have students share their estimates and reasoning when solving for some of the answers.

Activity

▲ Divvy Up

Materials: Divvy Up Game (BLM), Number Cards (BLM) 0–9, four sets

Players take turns drawing three number cards and creating a three-digit number with the cards. They write their numbers on one of the four division problem lines on a Divvy Up Game (BLM).

Players then complete the division problem, recording the quotient and remainder (if any). After four rounds, players total up their quotients and remainders and add them together.

The player with the greatest total score is the winner.

Alternatively, the player with the least total score wins.

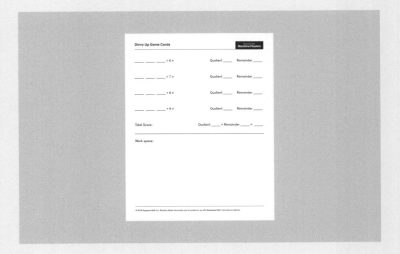

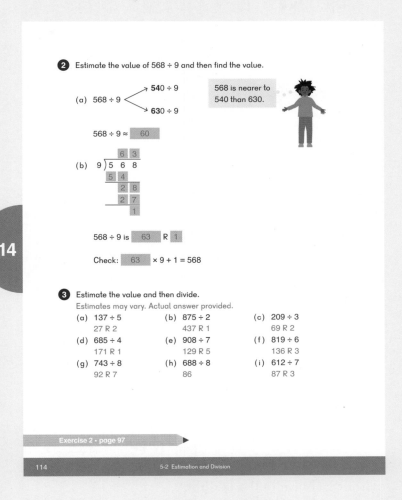

Lesson 3 Dividing 4-Digit Numbers

Objective
- Divide up to a four-digit number by a one-digit number.

Lesson Materials
- Place-value discs

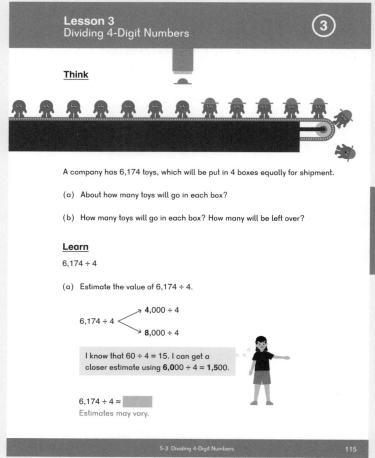

Think

Provide students with place-value discs and pose the **Think** problem. Have students estimate and then use the discs to find the number of toys that will go into each box.

Discuss student estimates and ideas for finding the actual amount.

Learn

Discuss the estimation methods shown in **Learn** and have students compare their own methods with the methods shown in the textbook.

Mei uses a multiple of 4 by thinking about 6,174 as 6,000. She knows 60 hundreds is a multiple of 4.

Ask students what is alike and what is different about dividing a three-digit number and dividing a four-digit number.

Work through the **Think** problem as demonstrated in **Learn**.

Have students relate the numbers in the algorithm with the steps shown in **Learn** to the place-value discs.

Ask students questions, such as:

- "What does the 1 in the thousands place of the quotient mean? Where can we see it with the place-value discs?"
- "Why did we subtract 4,000 from 6,000?"
- "Where is the remaining 2,000, that was not divided, recorded in the diagram?"

Emma checks her work with multiplication and adds in the remainder.

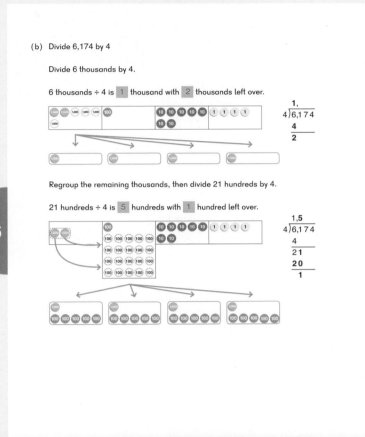

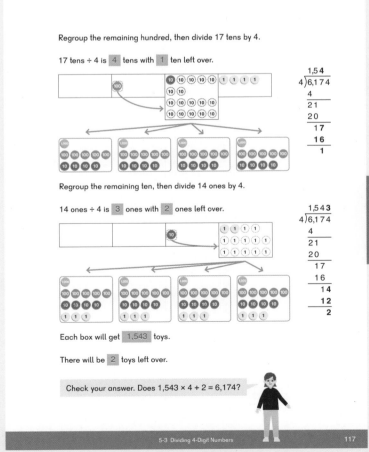

Do

1–**2** Students who need additional help should solve the problem with place-value discs first, then relate the steps with the discs to the steps shown in the textbook.

1 Dion uses 78 as a multiple of 6 to estimate 7,800 ÷ 6. 7,800 is closer to 8,092 than 6,000.

2 Students should see that since 2,408 is less than 3,000, there will not be a digit in the thousands column. The 2 thousands are regrouped into 24 hundreds.

The answer will be a three-digit number.

Alex reminds students how to record the quotient when a given place in the quotient has the value of 0.

Skipping the step of placing a 0 in the quotient where needed is a common student mistake. This is one reason it is so important for students to first estimate, then check for reasonableness after calculating.

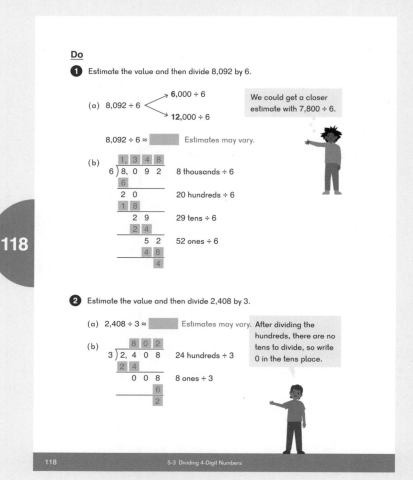

5 Students should be working with just the algorithm.

Students may see problems that can easily be solved using mental math.

For example:

(b)
$$3{,}248 \div 8 = 400 + 6$$
$$\diagup \quad \diagdown$$
$$3{,}200 \quad 48$$

(f)
$$\text{Dion knows } 4{,}018 \div 2 = 2{,}000 + 9$$
$$\diagup \quad \diagdown$$
$$4{,}000 \quad 18$$

Activity

▲ Divvy Up

Materials: Divvy Up Game (BLM), Number Cards (BLM) 0–9, four sets

Extend this game from the previous lesson to four-digit numbers. Players take turns drawing four number cards and creating a four-digit number. They write their numbers on one of the four division problem lines on a Divvy Up Game (BLM).

Players then complete the division problem, recording the quotient and remainder (if any). After four rounds, players total up their quotients and remainders and add them together.

The player with the greatest total score is the winner.

Alternatively, the player with the least total score wins.

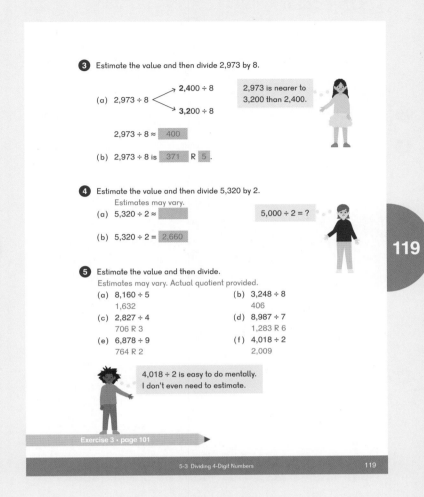

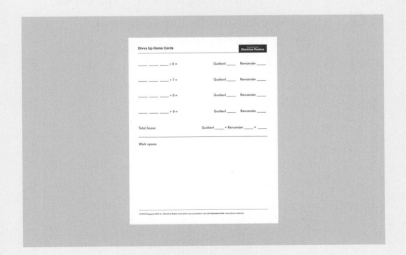

Exercise 3 • page 101

Lesson 4 Practice A

Objective

- Practice division.

1 Ask students how they solved the problems mentally.

2 Have students share their estimates for some of the problems.

3
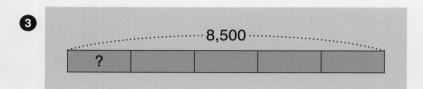

5 units ⟶ 8,500
1 unit ⟶ 8,500 ÷ 5 = 1,700

4

This problem asks: "How many $6 are in $580?"

4 – **5** show different interpretations of a remainder. In **4**, the money will buy 96 compasses and $4 will remain. In **5**, to raise $500 while the remainder means part of a hat, a full extra hat must be sold. 83 hats is not enough. Dion must sell 84 hats.

Lesson 4 Practice A

1 Divide.

(a) 800 ÷ 4   (b) 960 ÷ 3   (c) 450 ÷ 5
    200   320   90
(d) 840 ÷ 7   (e) 2,700 ÷ 3   (f) 9,200 ÷ 2
    120   900   4,600
(g) 7,000 ÷ 5   (h) 7,800 ÷ 6   (i) 81,000 ÷ 3
    1,400   1,300   27,000

2 Estimate the value and then divide.
Estimates may vary. Actual answer provided.

(a) 734 ÷ 5   (b) 851 ÷ 4   (c) 282 ÷ 6
    146 R 4   212 R 3   47
(d) 364 ÷ 8   (e) 5,047 ÷ 3   (f) 9,528 ÷ 8
    45 R 4   1,682 R 1   1,191
(g) 6,642 ÷ 7   (h) 9,975 ÷ 2   (i) 3,604 ÷ 9
    948 R 6   4,987 R 1   400 R 4

3 Karen wants to save $8,500 in 5 years. She wants to save the same amount each year. How much does she need to save each year?
8,500 ÷ 5 = 1,700
$1,700 each year

4 A teacher has $580 to spend on compasses. Each compass costs $6.

(a) Estimate the number of compasses she can buy.
600 ÷ 6 = 100; about 100 compasses
(b) Find the greatest number of compasses she can buy.
580 ÷ 6 is 96 R 4; 96 compasses
(c) How much money will she have left over if she buys that many?
$4 left over

5 A similar model to that in 4 can be drawn.

This problem asks, "How many $6 are in $500?"

7
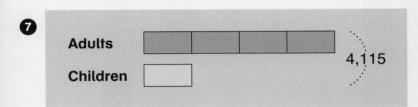

8
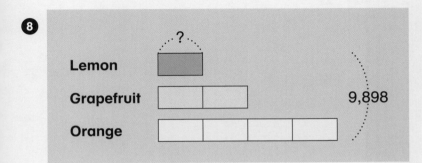

Activity

▲ Greatest and Least

Materials: Number Cards (BLM) 1–9

Using five of the digits 1–9 once, what is the greatest quotient that can be made? What is the least?

What if the divisor is 3?

Exercise 4 • page 105

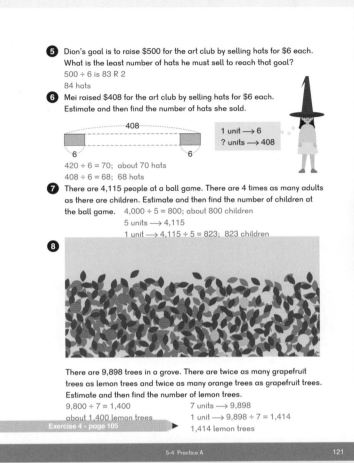

© 2019 Singapore Math Inc.   Teacher's Guide 4A Chapter 5   157

Lesson 5 Word Problems

Objective

- Solve multi-step word problems involving multiplicative comparison.

Think

Pose the **Think** problem and discuss Alex's questions. Students should think about which of the friends' number of pieces of trash could be used as the unit. If needed, have students redraw the model so that they can mark the model up further.

Prompt students by asking:

- "What information is given on the model?"
- "How can we make the bars for each friend's number of pieces of trash the same length?"
- "Which bar could we consider as the unit?"

Discuss student solutions.

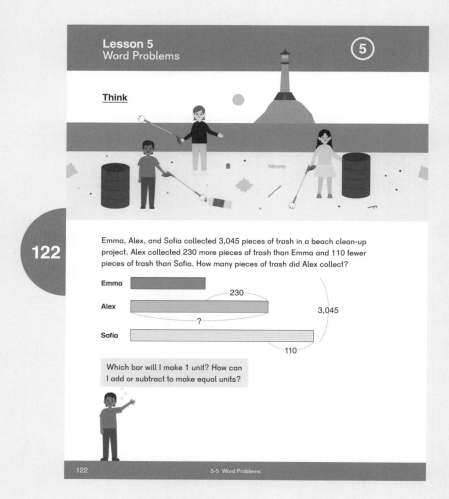

Learn

Discuss the methods shown in **Learn** and have students compare their own methods with the methods shown in the textbook. By making all of the bars the same length, each bar is an equal unit, and the value of each unit can be found using division, since the total is known.

Method 1

If Emma had found 230 more and Sofia had found 110 less, the three friends would all have the same amount, and there would be 3 equal units. One of those units represents the number of Alex's pieces of trash.

If Alex's bar represents the unit:

Emma = Alex − 230
Sofia = Alex + 110

Alex + Emma + Sofia = 3 units

3 units ⟶ 3,045 + 230 − 110 = 3,165

By dividing by 3, solve for 1 unit and find the number of Alex's pieces of trash.

Method 2

If Alex had found 230 less and Sofia had found 110 + 230 less, the three friends would all have the same amount, and there would be 3 equal units.

One of those units represents the number of Emma's pieces of trash.

If Emma's bar represents the unit:

Alex = Emma + 230
Sofia = Emma + 230 + 110
Emma + Alex + Sofia = 3 units

3 units ⟶ 3,045 − 230 − 230 − 110 = 2,475

We subtract to find the value of 3 units, or 2,475. By dividing by 3, we get the number of Emma's pieces of trash.

Add the number of Emma's pieces of trash (825) and 230 together to get the number of Alex's pieces of trash.

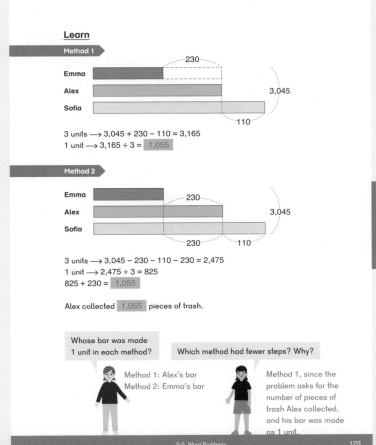

Do

① By first subtracting the amount of cards Dion kept for himself, students can find the value of the remaining 5 equal units and solve for 1 unit.

② 1 unit is the cost of one chair.

6 units + 1 unit + 284 = 1,299
7 units ⟶ 1,299 − 284 = 1,015

③ If the cost of the table lamp is considered one unit, then the cost of the two floor lamps are each 1 unit minus 70.

To make the cost of floor lamps equal to a unit, we add 70 to each floor lamp, which adds $140 to the total amount Mr. Lopez spent, which is:

3 units + (1 unit + 70) + (1 unit + 70)
5 units ⟶ 505 + 140 = 645

Divide the value of 5 units, or 645, by 5 to find the value of each table lamp unit:

5 units ⟶ 645
1 unit ⟶ 645 ÷ 5 = 129

To find the total cost of three table lamps:

1 unit ⟶ 129
3 units ⟶ 129 × 3 = 387

Note that students could solve the problem similarly to **Method 2** in **Learn** and use one floor lamp as the unit.

5 units ⟶ 505 − (3 × 70) = 295
1 unit ⟶ 295 ÷ 5 = 59

One floor lamp costs $59.
One table lamp costs $129 (59 + 70 = 129).
Three table lamps cost $387 (3 × 129 = 387).

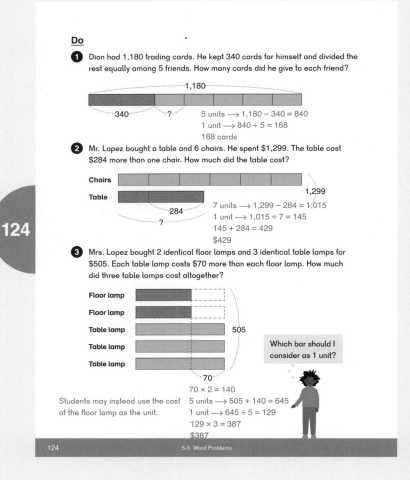

④ If students are confused about how the units were found, have them redraw the model step by step.

When redrawing the bar model, represent the number of books in the first and second boxes. The first box has 25 more books than the second box.

Add another bar representing the number of books in the third box. Students should think about what "twice as many" means in relation to the third box. That is, the bar for the third box is half as long as the second box:

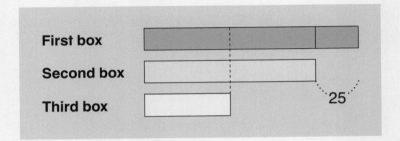

5 equal units + 25 books is 430 books in all.
Subtract 25 from the total of 430 to find the value of 5 units.

5 units ⟶ 430 − 25 = 405
1 unit ⟶ 405 ÷ 5 = 81
2 units ⟶ 2 × 81 = 162

This thinking will help students interpret ⑤ as well.

⑤ 5 equal units + 75 jars is 980 jars of jam in all.
Subtract 75 from 980 to find the value of 5 equal units.

5 units ⟶ 980 − 75 = 905
1 unit ⟶ 905 ÷ 5 = 181
3 units ⟶ 3 × 181 = 543

⑥ Suggest students draw the second number as twice the length of the first number:

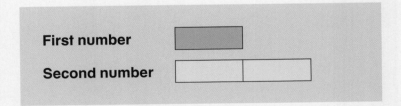

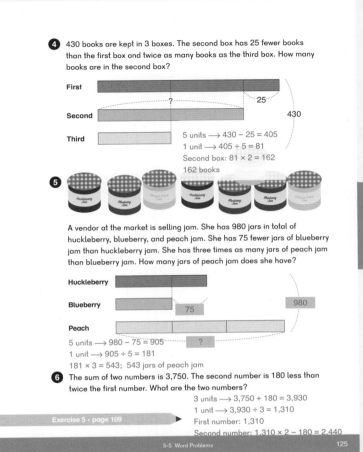

They can then show the 180 less than twice the first number on the model:

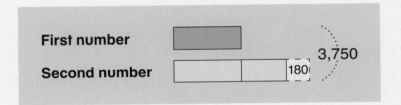

Add 180 to 3,750 to find the value of 3 equal units.

3 units ⟶ 3,750 + 180 = 3,930
1 unit ⟶ 3,930 ÷ 3 = 1,310
2 units ⟶ 1,310 × 2 = 2,620
Second number: 2,620 − 180 = 2,440

Exercise 5 • page 109

Lesson 6 Challenging Word Problems

Objective

- Extension: Solve multi-step word problems.

Think

Pose the **Think** problem and discuss Emma's suggestion. Students should think about how to represent the number of tablets and notebooks before some tablets were sold and after some tablets were sold.

If students need prompting, ask:

- "Does the number of laptops change?"
- "What should we use to represent one unit, laptops or tablets?" (Laptops, because we are told there are 3 times as many tablets as laptops.)

Learn

Students should recognize that the phrase, "after 2,600 tablets were sold, there were 500 more tablets than laptops," means that they will still have more tablets than laptops. The bar for tablets will still be longer than the bar for laptops. Since the number of laptops does not change, they can relate the information in the drawings to the before and after models to see that:

2,600 + 500 is the same as two units of tablets.

As Alex suggests, students may also find the value of 1 unit first:

2 units ⟶ 3,100
1 unit ⟶ 3,100 ÷ 2 = 1,550
Before: 4 units ⟶ 1,550 × 4 = 6,200
After: 6,200 − 2,600 = 3,600

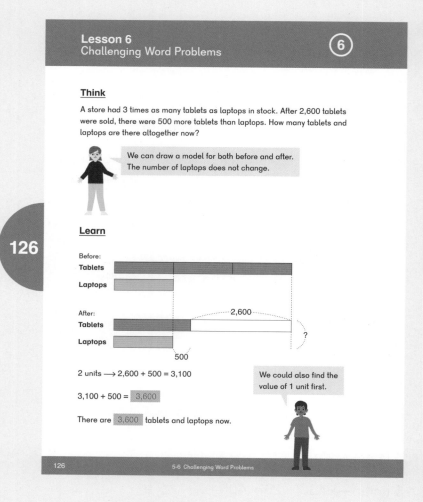

Do

1 If students need additional help determining how the units were found, have them redraw the model, step by step.

Dion suggests starting with the After model. Students can easily draw the number of vegan hot dogs to turkey hot dogs as 5 times as many. Then they can think about what the model looks like after 355 turkey hot dogs are sold.

Students can think about how to draw the bars. For students who need additional help, discuss the problem without numbers:

A vendor had turkey and vegan hot dogs.
She sold some of the turkey hot dogs.
After she sold the turkey hot dogs, she had 5 times as many vegan as turkey hot dogs.

Ask students, "How can we draw this?"

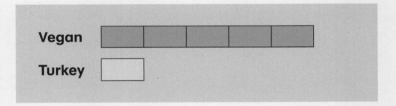

To work backwards, we know the number of vegan hot dogs does not change, so in the Before model, the vegan bar remains the same and the turkey bar increases by 355. It is hard to know how long to draw the part of the bar that represents 355 turkey hot dogs, so omit if needed:

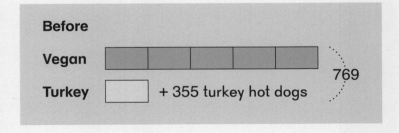

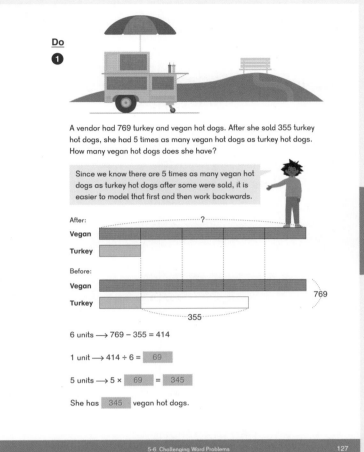

This can help students see that 6 units + 355 = 769. To find the value of 6 units, subtract 355 from 769:

6 units ⟶ 769 − 355 = 414
1 unit ⟶ 414 ÷ 6 = 69
5 units ⟶ 5 × 69 = 345
She has 345 vegan hot dogs.

❷ Sofia shares how to find the value of 1 unit of shoes before some were sold. From the model, we can see that:

Before, there are 3 units of blue shoes and 1 unit of red shoes.

After some of the blue and red shoes are sold, the number of blue shoes decreases by 360. We do not know how many blue shoes there are.

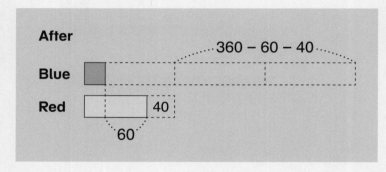

To find the difference between the two colors of shoes in the After model, we can see that 2 units is the same value as 360 − 60 − 40.

2 units ⟶ 360 − 60 − 40 = 260
1 unit ⟶ 260 ÷ 2 = 130

Next, to find the number of blue shoes there were at first, find the value of 3 units:

3 units ⟶ 3 × 130 = 390

Finally, subtract the number of blue shoes sold from the amount in the Before model.

390 − 360 = 30

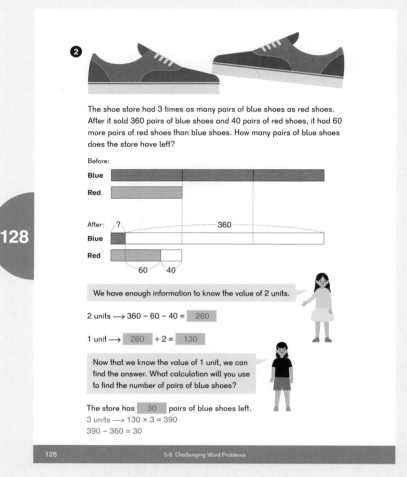

❸ — ❹ These problems are related and encourage students to think about how a bar model can help them solve such problems.

❸ People are moving out of North Town, but not out of South Town. The value for South Town is the same for Before and After.

The After bar for North Town shows that 2 units and some more represents 1,800 people. We can compare that to the Before model and see that 1,800 + 360 is the value of the 3 unit difference between North Town's population before and after.

3 units ⟶ 1,800 + 360 = 2,160
1 unit ⟶ 2,600 ÷ 3 = 720

There are 720 people living in South Town.

❹ People have moved to East Town and there are 360 more residents.

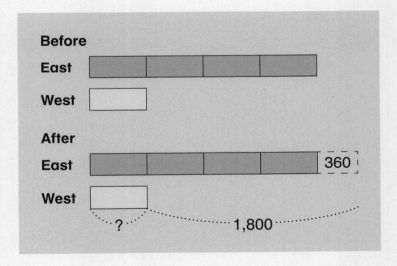

People are moving to East Town, but not to West Town. The value for West Town is the same in the Before and After models.

In the After model, we see that 3 units + 360 = 1,800.

3 units ⟶ 1,800 − 360 = 1,440
1 unit ⟶ 1,440 ÷ 3 = 480

There are 480 people living in West Town.

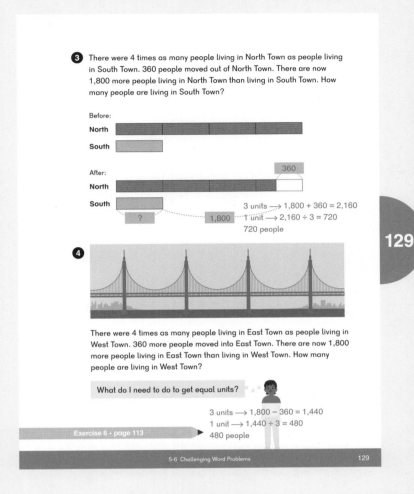

Exercise 6 • page 113

Lesson 7 Practice B

Objective

- Practice division.

1 Have students share their estimates for solving some of the problems.

2

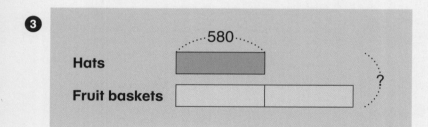

3

Hats ——580——
Fruit baskets ——?——

Lesson 7
Practice B   P 7

1 Estimate the value and then divide.
Estimates may vary. Actual answers provided.
(a) 602 ÷ 3   (b) 785 ÷ 9   (c) 518 ÷ 5
   200 R 2   87 R 2   103 R 3
(d) 6,287 ÷ 2   (e) 2,086 ÷ 7   (f) 4,905 ÷ 8
   3,143 R 1   298   613 R 1
(g) 5,970 ÷ 4   (h) 8,004 ÷ 6   (i) 9,986 ÷ 8
   1,492 R 2   1,334   1,248 R 2

130

2

Connor has twice as many trading cards as Hudson. Jamal has 3 times as many trading cards as Connor. Altogether, Hudson and Jamal have 595 cards.
       7 units ⟶ 595
(a) How many trading cards does Connor have?   1 unit ⟶ 595 ÷ 7 = 85
   85 × 2 = 170; 170 cards
(b) How many more trading cards does Jamal have than Connor?
   85 × 4 = 340; 340 more cards
(c) How many trading cards do all three boys have altogether?
   85 × 9 = 765 cards   or: 595 + 170 = 765
   765 cards altogether

3 Julia raised $580 selling hats. She raised twice as much money selling fruit baskets. She gave the money equally to 5 different charities. How much money did she give to each charity?
1 unit ⟶ 580
3 units ⟶ 3 × 580 = 1,740
1,740 ÷ 5 = 348; $348

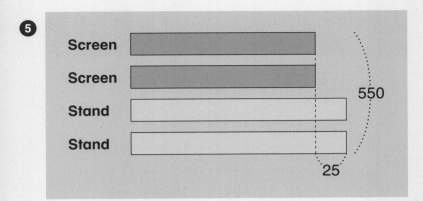

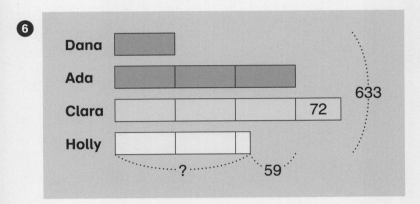

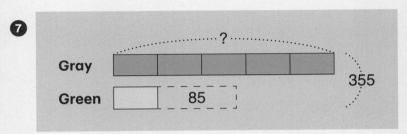

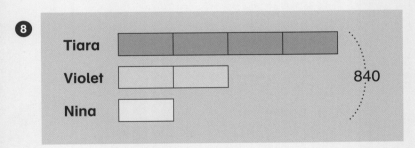

Exercise 7 • page 118

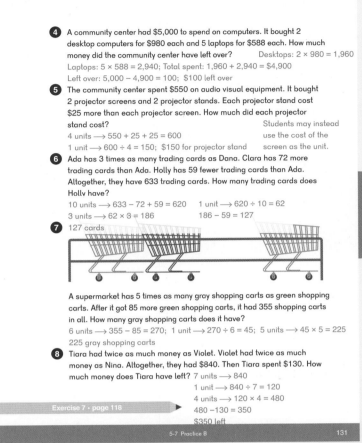

- When she packs 6 toy cars in each box she has 1 toy car left over.
- When she packs 5 toy cars in each box she has 1 toy car left over.
- When she packs 4 toy cars in each box, she has 1 toy car left over.
- When she packs 3 toy cars in each box, she has 1 toy car left over.

What is the smallest possible number of toy cars Mei could have?

Mei could have 61 cars:

$61 \div 6 = 10$ R 1
$61 \div 5 = 12$ R 1
$61 \div 4 = 15$ R 1
$61 \div 3 = 20$ R 1

Brain Works

★ **Toys**

Mei has toy cars that she is putting into boxes.

Review 1

Objective

- Cumulative review of content from Chapters 1–5.

Use the cumulative review to practice and reinforce content and skills from the first five chapters.

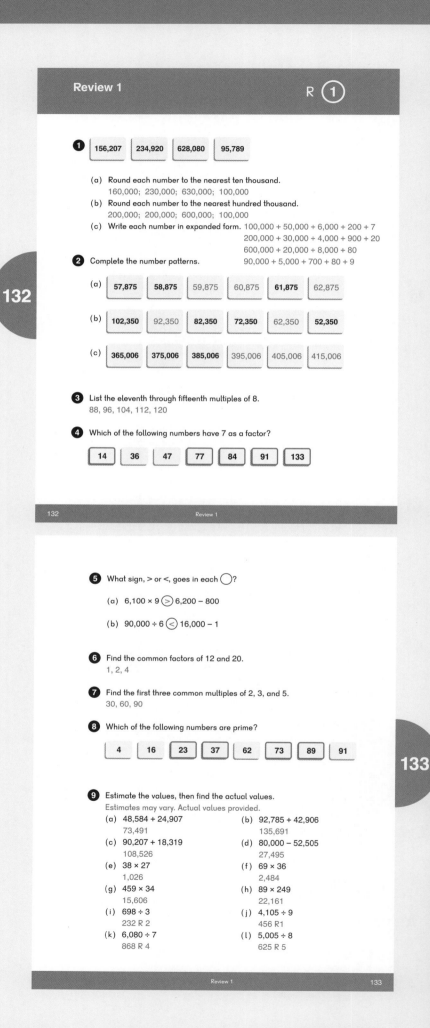

⑩ Students must find common factors of 24, 32 and 40. The common factors are 1, 2, 4, and 8.

⑪ Students must find the first common multiple of 6, 10, and 15, which is 30.

⑫ 340 × 24 weeks = 8,160
Madison earns $8,160.

8,160 − 2,000 saved − 3,360 tuition = 2,800

She spent $2,800 on the 2 couches.

⑬ 27 buses × 24 children = 648 children on buses.
700 children − 648 children on buses = 52 children left to be transported.

52 ÷ 4 minibuses = 13 children on each minibus.

⑭ Students may not need to draw a Before and After model to solve the problem. Have them begin by drawing the Before model.

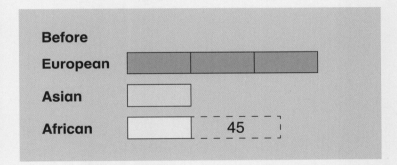

Next, they can simply add the rest of the information: 55 more Asian coins and the total number of coins.

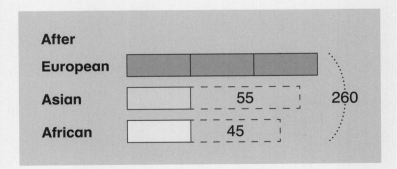

From this model we can see:

5 units ⟶ 260 − 55 − 45 = 160

1 unit ⟶ 160 ÷ 5 = 32

European coins: 3 units ⟶ 3 × 32 = 96

Asian coins: 32 + 55 = 87

African coins: 32 + 45 = 77

Exercise 8 • page 121

Exercise 1 • pages 95–96

Chapter 5 Division

Exercise 1

Basics

1. Use mental calculation to find the quotients.

   (a) 63 ÷ 3 = 21
       60 3

   (b) 72 ÷ 3 = 24
       60 12

   (c) 55 ÷ 5 = 11
       50 5

   (d) 75 ÷ 5 = 15
       50 25

   (e) 48 ÷ 4 = 12
       40 8

   (f) 68 ÷ 4 = 17
       40 28

2. (a) 72 ÷ 6 = 12
   (b) 72 tens ÷ 6 = 12 tens = 120
   (c) 72 hundreds ÷ 6 = 12 hundreds = 1,200
   (d) 72 thousands ÷ 6 = 12 thousands = 12,000

3. (a) 84 ÷ 7 = 12
   (b) 840 ÷ 7 = 120
   (c) 8,400 ÷ 7 = 1,200
   (d) 84,000 ÷ 7 = 12,000

Practice

4. Use mental calculation to find the quotients.

   (a) 40,000 ÷ 8 = 5,000
   (b) 1,500 ÷ 5 = 300
   (c) 3,600 ÷ 4 = 900
   (d) 7,200 ÷ 9 = 800
   (e) 2,400 ÷ 8 = 300
   (f) 2,700 ÷ 3 = 900
   (g) 28,000 ÷ 7 = 4,000
   (h) 6,400 ÷ 8 = 800
   (i) 45,000 ÷ 3 = 15,000
   (j) 100 ÷ 5 = 20
   (k) 91,000 ÷ 7 = 13,000
   (l) 72,000 ÷ 4 = 18,000
   (m) 10,800 ÷ 9 = 1,200
   (n) 1,500 ÷ 2 = 750

5. A couch costs $360. It costs twice as much as a table. A table costs three times as much as a chair. How much does a chair cost?
   1 unit ⟶ 360 ÷ 6 = 60
   $60

6. A factory produced 90,000 jars of jam in 6 hours. If it produces the same number of jars each hour, how many jars does it produce in 2 hours?
   1 hour: 90,000 ÷ 6 = 15,000
   2 hours: 15,000 × 2 = 30,000
   30,000 jars of jam

Exercise 2 • pages 97–100

Exercise 2

Basics

1 (a) Complete the following estimates for the quotient of 743 ÷ 3.

$600 ÷ 3 = $ 200

$750 ÷ 3 = $ 250

$900 ÷ 3 = $ 300

(b) Divide 743 by 3.

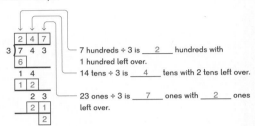

7 hundreds ÷ 3 is __2__ hundreds with 1 hundred left over.

14 tens ÷ 3 is __4__ tens with 2 tens left over.

23 ones ÷ 3 is __7__ ones with __2__ ones left over.

743 ÷ 3 is __247__ with a remainder of __2__.

(c) Compare the estimates to the actual quotient. Which estimate(s) were lower? Which estimate(s) were higher? Which estimate was closest?
Lower: 200; Higher: 250, 300
Closest: 250

(d) Check: __247__ × 3 + __2__ = 743

Practice

2 (a) Emma's estimate: 878 ÷ 9 ≈ 900 ÷ 9
Alex's estimate: 878 ÷ 9 ≈ 880 ÷ 10
Mei's estimate: 878 ÷ 9 ≈ 810 ÷ 9

List the estimates in order from least to greatest.
Alex's estimate: 88
Mei's estimate: 90
Emma's estimate: 100

(b) Divide 878 by 9.

```
      9 7
9)8 7 8
  8 1
    6 8
    6 3
      5
```

97 with a remainder of 5

3 Which of the following gives the greatest quotient? Circle it.

653 ÷ 7   700 ÷ 8   498 ÷ 5

4 A community boating center had $200. It bought 6 identical life vests for its customers to use and had $8 left. Using estimation, circle the most reasonable cost for each life vest.

$21 $32 $48

5 (a) Estimate and then divide.   Estimates may vary. Actual quotients provided.

848 ÷ 3 ≈ ☐   972 ÷ 4 ≈ ☐   715 ÷ 2 ≈ ☐

```
    2 8 2          2 4 3          3 5 7
3)8 4 8        4)9 7 2        2)7 1 5
  6              8              6
  2 4            1 7            1 1
  2 4            1 6            1 0
    0 8            1 2            1 5
      6            1 2            1 4
      2            0              1
```

625 ÷ 7 ≈ ☐   556 ÷ 8 ≈ ☐   828 ÷ 4 ≈ ☐

```
    8 9              6 9          2 0 7
7)6 2 5        8)5 5 6        4)8 2 8
  5 6            4 8            8
    6 5            7 6            0 2 8
    6 3            7 2            2 8
      2            4              0
```

(b) Put the quotients in order from least to greatest. Add the middle two numbers. Divide that answer by the sum of the remainders. You will get 50 if you did all the calculations correctly.
69 R 4, 89 R 2, 207, 243, 282 R 2, 357 R 1
4 + 2 + 2 + 1 = 9
207 + 243 = 450
450 ÷ 9 = 50

6 At a fair, balloons were given out in this order: yellow, blue, blue, red, red, red, yellow, blue, blue, red, red, red, and so on until 250 balloons were given out.

Y B B R R R Y B B R R R

(a) What color was the 250th balloon? Red
250 ÷ 6 is 41 R 4
The 41st is the end of one unit, so the 4th one after that is red.

(b) How many red balloons were given out? 124
There are 3 red balloons in each set of 6.
Then 1 additional red balloon was given out.
41 × 3 + 1 = 124

Challenge

7 (a) Find the remainder of 68 ÷ 9.
5

Find the remainder of 69 ÷ 9.
6

Find the remainder of the sum of the two remainders divided by 9.
2

Find the remainder of 68 + 69 divided by 9.
2

What do you notice? Try it with some other numbers.
The two remainders, from the last two steps, are the same.
Answers will vary.

(b) Find the remainder when 17 + 15 + 11 is divided by 4.
Find the sum of the remainders when each of the addends is divided by 4. Then find the remainder when that sum is divided by 4.
1 + 3 + 3 = 7; 7 ÷ 4 is 1 with a remainder of 3.
3

(c) Find the remainder when 69 + 69 + 74 + 58 is divided by 8.
5 + 5 + 2 + 2 = 14
14 ÷ 8 is 1 with a remainder of 6.
6

Teacher's Guide 4A Chapter 5

Exercise 3 • pages 101–104

Exercise 3

Basics

1 (a) Complete the following estimations for the quotient of 9,875 ÷ 4.

8,000 ÷ 4 = 2,000
10,000 ÷ 4 = 2,500
12,000 ÷ 4 = 3,000

(b) Divide 9,875 by 4.

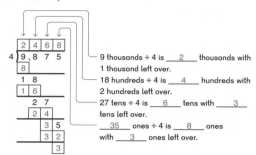

9 thousands ÷ 4 is __2__ thousands with 1 thousand left over.
18 hundreds ÷ 4 is __4__ hundreds with 2 hundreds left over.
27 tens ÷ 4 is __6__ tens with __3__ tens left over.
__35__ ones ÷ 4 is __8__ ones with __3__ ones left over.

9,875 ÷ 4 is __2,468__ with a remainder of __3__.

(c) Compare the estimates to the actual quotient. Which estimate(s) were lower? Which estimate(s) were higher? Which estimate was closest?
Lower: 2,000; Higher: 2,500, 3,000; Closest: 2,500

(d) Check: 2,468 × 4 + 3 = 9,875

Practice

2 (a) Sofia's estimate: 4,975 ÷ 3 ≈ 3,000 ÷ 3
Mei's estimate: 4,975 ÷ 3 ≈ 6,000 ÷ 3
Dion's estimate: 4,975 ÷ 3 ≈ 4,500 ÷ 3

List the estimates in order from least to greatest.
Sofia's estimate: 1,000, Dion's estimate: 1,500, Mei's estimate: 2,000

(b) Divide 4,975 by 3.

```
      1 6 5 8
   3)4 9 7 5
     3
     1 9
     1 8
       1 7
       1 5
         2 5
         2 4
           1
```
1,658 R 1

3 6,798 ÷ 5 is closest to which number? Circle it.

| 150 | 1,000 | **1,500** | 2,000 |

4 Two of these expressions have the same quotient. Circle them.

| 4,950 ÷ 6 | **6,531 ÷ 7** | 3,780 ÷ 5 | **2,799 ÷ 3** |

5 Is the quotient of 7,489 ÷ 8 three or four digits?
Three digits

---

6 (a) Estimate and then divide. *Estimates may vary. Actual quotients provided.*

8,375 ÷ 6 ≈

9,971 ÷ 2 ≈

7,761 ÷ 9 ≈

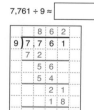

3,945 ÷ 5 ≈

(b) Add the quotients and the remainders together, then divide that sum by 5. You will get 1,608 if you did all the calculations correctly.
1,395 + 4,985 + 862 + 789 + 5 + 1 + 3 = 8,040
8,040 ÷ 5 = 1,608

7 How many egg cartons are needed for 2,560 eggs if each carton holds 6 eggs?
2,560 ÷ 6 is 426 R 4
427 cartons

8 Students are seated in the following order on a train: 1= left window, 2 = left aisle, 3 = right aisle, 4 = right window, 5 = left window, etc. Jacob gets seat 1,423. Where is his seat?

| left window | left aisle | **right aisle** | right window |

1,423 ÷ 4 is 355 R 3

Challenge

9 Find the missing numbers.

(a)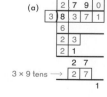

3 × 9 tens →

(b)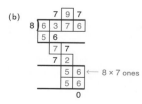

← 8 × 7 ones

Exercise 4 • pages 105–108

Exercise 4

Check

1. Use mental calculation to find the quotients.

   (a) 2,000 ÷ 5 = **400**   (b) 560 ÷ 8 = **70**

   (c) 90,000 ÷ 6 = **15,000**   (d) 15,000 ÷ 2 = **7,500**

   (e) 6,400 ÷ 4 = **1,600**   (f) 75,000 ÷ 3 = **25,000**

2. Circle the number that is equal to 6,461 ÷ 7. Use estimation.

   795   893   **923**   1,373

3. Two of these expressions have the same quotient. Circle them.

   598 ÷ 6   **1,400 ÷ 8**   **875 ÷ 5**   4,007 ÷ 3

4. Write > or < in each ○. Use estimation.

   (a) 8,268 ÷ 4 **>** 2,000   (b) 842 ÷ 7 **<** 276 ÷ 2

   (c) 3,824 ÷ 9 **<** 4,000 ÷ 8   (d) 5,723 ÷ 6 **<** 4,789 ÷ 4

   (e) 9,199 ÷ 3 **>** 587 × 4   (f) 7,155 ÷ 8 **<** 528 + 679

5. (a) Divide.

A: 8,989 ÷ 8 = 1,123 R 5	B: 5,693 ÷ 7 = 813 R 2
C: 6,725 ÷ 5 = 1,345	D: 5,301 ÷ 9 = 589
E: 8,655 ÷ 6 = 1,442 R 3	F: 6,754 ÷ 2 = 3,377

   (b) Write the digits of the answers and the remainders from problems A to F in order below. Add consecutive numbers together. If you did all the calculations correctly, you will discover a pattern.

   1, 1, 2, 3, 5, 8, 1 3, 2 1, 3 4,
   5 5, 8 9, 1 4 4, 2 3 3, 3 7 7 …

   The sum of consecutive numbers is the next number.

6. 6 kayak paddles cost $348. What is the cost of 1 kayak paddle?

   348 ÷ 6 = 58
   $58

7. A 5-person pedal boat costs 5 times as much as a stand-up paddle board. If the pedal boat costs $1,245, how much does the paddle board cost?

   1 unit → 1,245 ÷ 5 = 249
   $249

8. A community boating center had $2,000. It bought 8 identical kayaks and had $8 left. How much did each kayak cost?

   2,000 − 8 = 1,992
   1,992 ÷ 8 = 249
   $249

Challenge

9. Mei plays a game with coins. She puts 4 coins in a row, heads up. In the first move, she turns over the first coin. In the second move, she turns over the first two coins. In the third move, she turns over the first three coins. In the fourth move, she turns over all four coins. In the fifth move, she starts over again, turning over the first coin. And so on. What side of each coin is facing up after 1,995 moves?

	Coin 1	Coin 2	Coin 3	Coin 4
Starting position	H	H	H	H
1st move	T	H	H	H
2nd move	H	T	H	H
3rd move	T	H	T	H

   …

   Students can make a table and continue the pattern. At the end of the 8th move, the coins are all heads up again. The 9th move will be like the 1st move. The pattern repeats every 8 times. 1,995 ÷ 8 is 249 with a remainder of 3. So after the 1,995th move the positions will be the same as after the 3rd move: T, H, T, H.

Exercise 5 • pages 109–112

Exercise 5

Basics

1. There are 3 neighboring orchards, A, B, and C. Orchard A has 60 fewer fruit trees than Orchard B. Orchard C has 3 times as many fruit trees as Orchard B. If the three orchards have 430 fruit trees altogether, how many fruit trees does Orchard C have?

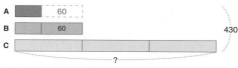

5 units ⟶ 430 + 60 = 490
1 unit ⟶ 490 ÷ 5 = 98
3 units ⟶ 3 × 98 = 294
294 fruit trees

2. There are 3 neighboring orchards, D, E, and F. Orchard D has 60 fewer fruit trees than Orchard E. Orchard D has 3 times as many fruit trees as Orchard F. If the three orchards have 424 fruit trees altogether, how many fruit trees does Orchard F have?

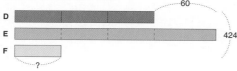

7 units ⟶ 424 − 60 = 364
1 unit ⟶ 364 ÷ 7 = 52
52 fruit trees

3. There are 3 neighboring orchards, K, L, and M. Orchard K has 4 times as many fruit trees as Orchard L. Orchard M has 60 fewer fruit trees than Orchard K. If Orchards K and L together have 430 fruit trees, how many fruit trees does Orchard M have?

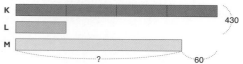

5 units ⟶ 430
1 unit ⟶ 430 ÷ 5 = 86
4 units ⟶ 4 × 86 = 344
Orchard M: 344 − 60 = 284
284 fruit trees

4. A paddle for adults costs $10 more than a paddle for kids. The boating center bought 4 of each type of paddle. The total cost was $312. What was the total cost for 4 adult-sized paddles?

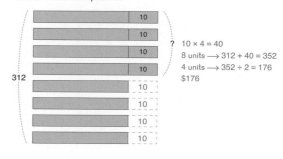

10 × 4 = 40
8 units ⟶ 312 + 40 = 352
4 units ⟶ 352 ÷ 2 = 176
$176

Practice Methods may vary.

5. Three siblings bought a small sailboat together. The sailboat cost $3,150. Ada contributed $20 more than Bron. Cora contributed $130 less than twice as much money as Ada. How much money did Bron contribute?

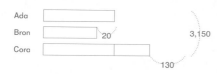

4 units ⟶ 3,150 + 130 + 20 = 3,300
1 unit ⟶ 3,300 ÷ 4 = 825
825 − 20 = 805
$805

6. A community boating center had $12,500. It bought 3 surf skis and 1 keelboat and had $265 left over. The keelboat cost $6,455 more than a surf ski. How much did the keelboat cost?

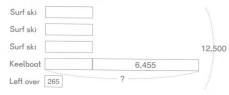

4 units ⟶ 12,500 − 265 − 6,455 = 5,780
1 unit ⟶ 5,780 ÷ 4 = 1,445
Keelboat: 1,445 + 6,455 = 7,900
$7,900

7. An apartment complex manager is replacing some of the appliances. He bought 3 ovens and 4 refrigerators for $4,583. Each oven cost $570 less than a refrigerator. What is the cost of one oven?

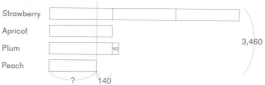

4 × 570 = 2,280
7 units ⟶ 4,583 − 2,280 = 2,303
1 unit ⟶ 2,303 ÷ 7 = 329
$329

8. A factory produced 3,460 jars of jam. It produced 3 times as many jars of strawberry jam as apricot jam, 60 more jars of plum jam than apricot jam, and 140 fewer jars of peach jam than apricot jam. How many jars of peach jam did it produce?

6 units ⟶ 3,460 − 60 + 140 = 3,540
1 unit ⟶ 3,540 ÷ 6 = 590
590 − 140 = 450
450 jars of peach jam

Exercise 6 • pages 113–117

Exercise 6

Basics

1 In a car sales lot there were three times as many vans as sedans. After 105 vans were sold, there were twice as many sedans as vans. How many cars were there in the end?

1 unit ⟶ 105 ÷ 5 = 21
3 units ⟶ 3 × 21 = 63
63 cars

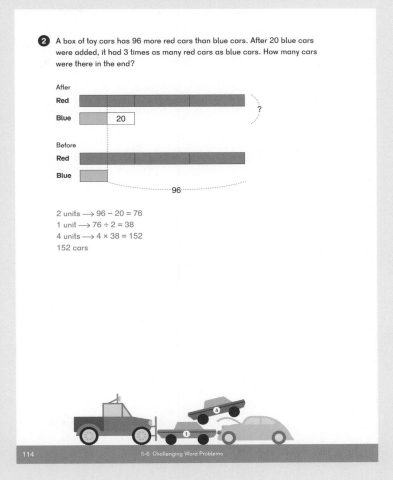

2 A box of toy cars has 96 more red cars than blue cars. After 20 blue cars were added, it had 3 times as many red cars as blue cars. How many cars were there in the end?

2 units ⟶ 96 − 20 = 76
1 unit ⟶ 76 ÷ 2 = 38
4 units ⟶ 4 × 38 = 152
152 cars

Practice

3 2 fish tanks, A and B, had a total of 196 fish. Tank A has more fish than Tank B. After 32 fish were added to Tank A and 15 fish in Tank B were sold, Tank A had 75 more fish than Tank B. How many fish did each tank have at first?

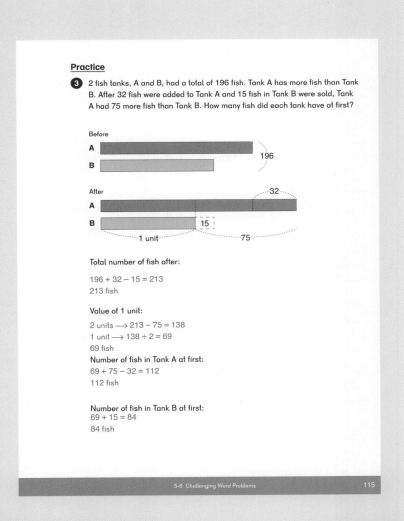

Total number of fish after:
196 + 32 − 15 = 213
213 fish

Value of 1 unit:
2 units ⟶ 213 − 75 = 138
1 unit ⟶ 138 ÷ 2 = 69
69 fish

Number of fish in Tank A at first:
69 + 75 − 32 = 112
112 fish

Number of fish in Tank B at first:
69 + 15 = 84
84 fish

4 Dorothy has $900 and Maria has $360. They each bought a couch at the same price. Now, Dorothy has 7 times as much money as Maria. How much did the couch cost?

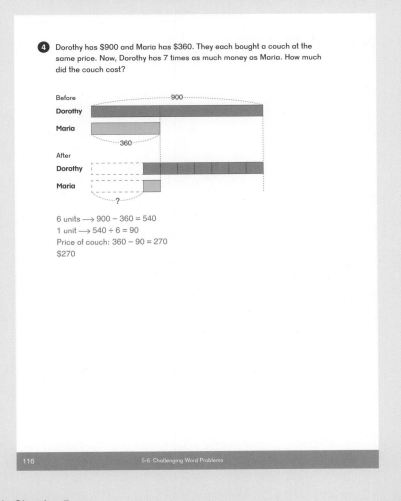

6 units ⟶ 900 − 360 = 540
1 unit ⟶ 540 ÷ 6 = 90
Price of couch: 360 − 90 = 270
$270

5 Daren saved 7 times as much money as Matt. They both saved another $370. Now they have $900 altogether. How much money has Daren saved?

Before
Daren
Matt

After
Daren 370
Matt 370
900

2 × 370 = 740
8 units ⟶ 900 − 740 = 160
1 unit ⟶ 160 ÷ 8 = 20
7 units ⟶ 7 × 20 = 140
140 + 370 = 510
$510

Exercise 7 • pages 118–120

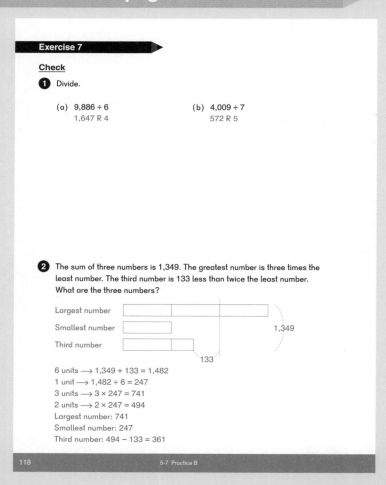

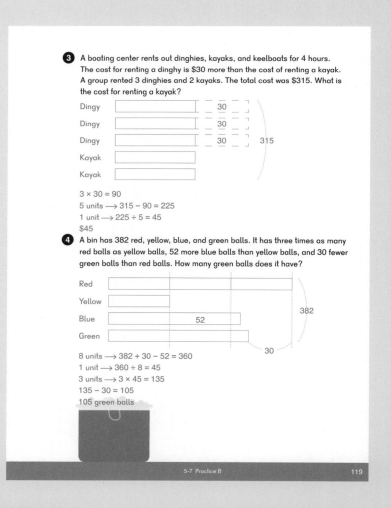

5. Cody wants to buy some apple trees for his property. He needs $48 more to buy 3 apple trees. After getting $240 more, he had enough money to buy 5 apple trees. How much does 1 apple tree cost?
Cost of last 2 apple trees: 240 − 48 = 192
Cost of 1 apple tree: 192 ÷ 2 = 96
$96

6. A number with the hundreds digit of 3 and tens digit of 2 is divided by 9. The remainder is 5. What is the ones digit of the number?
The largest product possible is 324 (since 329 minus the remainder of 5 is 324). Therefore, we should solve for the multiple of 9 that is less than or equal to 324.

36 × 9 = 324
324 + 5 = 329
Ones digit: 9

Exercise 8 • pages 121–126

Exercise 8

Check

1.

Number	Rounded to the nearest			
	100,000	10,000	1,000	100
134,710	100,000	130,000	135,000	134,700
634,550	600,000	630,000	635,000	634,600
98,432	100,000	100,000	98,000	98,400
250,500	300,000	250,000	251,000	250,500

2. The number 83,238 is a palindrome. It is the same read from left to right and from right to left (ignoring the comma). What is the 10th palindrome after 83,238?
83338, 83438, 83538, 83638, 83738, 83838, 83938,
84048, 84148, 84248
10th palindrome: 84,248

3. Use the clues to find the mystery 6-digit number.

   Clue 1 All the digits are different.
   Clue 2 The digit 9 is in the ten thousands place. Ten thousands place: 9
   Clue 3 The digit in the hundreds place is 6 less than the digit in the ten thousands place. Hundreds place: 9 − 6 = 3
   Clue 4 One of the digits stands for 5,000. Thousands place: 5
   Clue 5 The number is less than 200,000. Hundred thousands place: 1
   Clue 6 The digit in the tens place stands for 0. Tens place: 0
   Clue 7 The number is an odd number. Ones place: 7
   195,307

4. Use estimation to arrange the expressions in order from least to greatest.

   (a) A: 32 × 76   B: 846 + 1,158   C: 8,462 − 4,981   D: 178 × 8
   D, B, A, C

   (b) A: 16,982 + 33,174   B: 756 × 72   C: 409,563 − 343,669
   A, B, C

5. List all the common factors of 30 and 45.
   1, 3, 5, 15

6. List the first four common multiples of 3, 4, and 6.
   12, 24, 36, 48

© 2019 Singapore Math Inc.   Teacher's Guide 4A Chapter 5

7 (a) Can the sum of two prime numbers be an odd number? Explain why or why not.
Yes, but only if one of the numbers is 2. All prime numbers other than 2 are odd and the sum of two odd numbers is even.

(b) Can the product of two prime numbers be an odd number? Explain why or why not.
Yes, except for 2 all prime number are odd and the product of two odd numbers is odd.

8 A 2-digit odd number is a factor of 60 and a multiple of 3. What is the number?
15

9 Use the clues to determine which number, 61, 23, 72, or 51, is on the other side of the cards A, B, C, and D.

Clue 1 A prime number is between two composite numbers.
Clue 2 The odd multiple of 3 has no card to the right of it.
Clue 3 The least number is not between two cards.

A	B	C	D
23	72	61	51

61 and 23 are prime. The only odd multiple of 3 is 51.

10 A community boating center had 3 fundraisers last year for their youth sailing program. At the first fundraiser, they raised $18,945. At the second fundraiser, they raised $7,285 more than at the first fundraiser. At the third fundraiser, they raised $5,982 less than at the second fundraiser. How much money did they raise in all?
First fundraiser: 18,945
Second fundraiser: 18,945 + 7,285 = 26,230
Third fundraiser: 26,230 − 5,982 = 20,248
Total: 18,945 + 26,230 + 20,248 = 65,423
$65,423

11 A vendor at a farmers market is selling honey. He has 1,450 jars in total of alfalfa, blueberry, and clover honey. He has 130 more jars of alfalfa honey than blueberry honey. He has twice as many jars of clover honey as alfalfa honey. How many jars of alfalfa honey does he have?

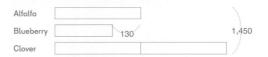

4 units ⟶ 1,450 + 130 = 1,580
1 unit ⟶ 1,580 ÷ 4 = 395
395 jars of alfalfa honey

12 3 ships leave a port on January 31. The first one returns to the port every 4 weeks, the second one every 8 weeks, and the third one every 10 weeks. In how many weeks will they all be in the port again?
The least common multiple of 4, 8, 10 is 40.
40 weeks

13 Oliver had 3 times as many rocks in his collection as Katie. Then Oliver collected 75 more rocks and Katie collected 105 more rocks. Now Katie has the same number of rocks as Oliver. How many rocks did Oliver have at first?

Before
Oliver
Katie

2 units ⟶ 105 − 75 = 30
1 unit ⟶ 30 ÷ 2 = 15
3 units ⟶ 15 × 3 = 45
45 rocks

After   1 unit
Oliver [75   ]
Katie [105   ]

Challenge

14 This year, John's age is a multiple of 3. Last year, his age was 1 less than a multiple of 4. Next year, his age will be 3 more than a multiple of 5. What is the youngest age he could be?
Multiples of 3: 3, 6, 9, 12, 15, 18, 21
Multiples of 4 − 1 + 1: 4, 8, 12, 16, 20
Multiples of 5 + 3 − 1: 7, 12, 17, 22
12 years old

15 In a textbook, 900 digits are used for the page numbers. How many pages are in the textbook, starting with page 1? (Hint: First find how many digits are used for pages 1-9 and 10-99.)
Pages 1- 9: 9 pages with 1 digit each. 9 × 1 = 9 digits
Pages 10 - 99: 90 pages with 2 digits each. 90 × 2 = 180 digits
Pages with 3 digits would be 100-999, assume for now there are fewer than 999 pages.
Number of digits on the remaining pages: 900 − 189 = 711
711 ÷ 3 = 237
99 + 237 = 336
336 pages

Chapter 6 Fractions

Overview

Suggested number of class periods: 8–9

Lesson		Page	Resources	Objectives
	Chapter Opener	p. 183	TB: p. 135	Review fraction concepts.
1	Equivalent Fractions	p. 184	TB: p. 136 WB: p. 127	Review equivalent fractions. Use multiplication and division to generate equivalent fractions.
2	Comparing and Ordering Fractions	p. 187	TB: p. 141 WB: p. 130	Compare fractions by finding common denominators or numerators.
3	Improper Fractions and Mixed Numbers	p. 190	TB: p. 145 WB: p. 133	Express fractions that are greater than 1 as improper fractions and mixed numbers.
4	Practice A	p. 192	TB: p. 148 WB: p. 136	Practice fraction concepts from the chapter.
5	Expressing an Improper Fraction as a Mixed Number	p. 194	TB: p. 150 WB: p. 139	Express improper fractions as mixed numbers.
6	Expressing a Mixed Number as an Improper Fraction	p. 196	TB: p. 152 WB: p. 142	Convert mixed numbers to improper fractions.
7	Fractions and Division	p. 198	TB: p. 154 WB: p. 145	Understand fractions as division.
8	Practice B	p. 202	TB: p. 159 WB: p. 148	Practice concepts from the chapter.
	Workbook Solutions	p. 203		

Chapter 6 Fractions

Notes

In Dimensions Math 3B students learned to:

- Compare and order fractions using common numerators, common denominators, and benchmark fractions.
- Locate fractions on a number line.
- Define a fraction as being composed by unit fractions.
- Find equivalent fractions.
- Express fractions greater than 1 as improper fractions.
- Express whole numbers equal to 1 and 2 as improper fractions (e.g. $\frac{3}{3} = 1, \frac{6}{3} = 2$).
- Add and subtract fractions with the same denominator.

This chapter reviews equivalent fractions and builds on that knowledge to formally introduce improper fractions and mixed numbers.

Equivalent Fractions

Fractions that have the same value are called equivalent fractions. Equivalent fractions are located at the same point on a number line.

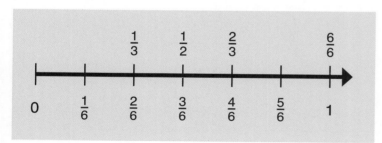

Equivalent fractions can be found by:

- Multiplying the numerator and denominator by the same whole number.

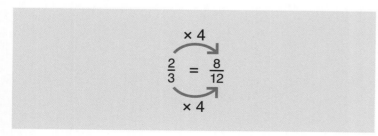

- Dividing the numerator and denominator by a common factor.

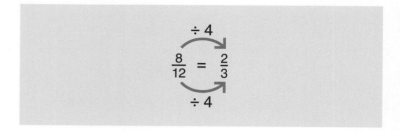

Comparing Fractions with Common Denominators

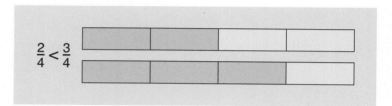

Because the numerator counts the number of equal parts, it is easy to compare fractions with the same denominator since you only have to compare the numerators.

Number lines also make it easy to compare fractions with the same denominator. A fraction on a number line is greater than the fractions to its left.

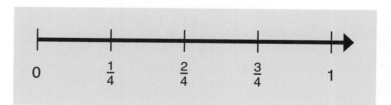

Chapter 6 Fractions

Notes

Comparing Fractions with Like Numerators

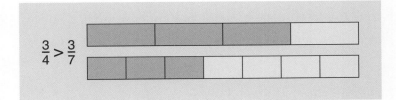

Students must understand that the unit fraction $\frac{1}{7}$ is less than the unit fraction $\frac{1}{4}$. If 1 is divided into 7 equal parts, each part is less than 1 divided into 4 equal parts. Since $\frac{1}{7}$ is less than $\frac{1}{4}$, $\frac{3}{7}$ will be less than $\frac{3}{4}$.

Through recognizing that the size of the part is smaller even though the denominator is greater, students develop critical understanding of denominators.

Comparing Fractions with Unlike Denominators

Students will apply their knowledge of multiples and factors to find an equivalent fraction when comparing fractions.

When neither numerators nor denominators are the same, students can calculate equivalent fractions to compare fractions with like denominators.

If asked to find which is greater, $\frac{3}{4}$ or $\frac{7}{10}$, students will find multiples of 4 and 10.

Multiples of 10: 10, **20**

Multiples of 4: 4, 8, 12, 16, **20**

When they find that 20 is a common multiple of 4 and 10, they will then calculate equivalent fractions to rewrite both fractions with a denominator of 20:

$\frac{3}{4} = \frac{15}{20}$

$\frac{7}{10} = \frac{14}{20}$

Since 15 > 14, $\frac{3}{4} > \frac{7}{10}$.

Using a common numerator can be helpful in comparing fractions where it is not easy to find common multiples of the denominator but it is easy to find common multiples of the numerator. For example, to compare $\frac{2}{9}$ and $\frac{4}{17}$, it is easier to compare $\frac{4}{18}$ and $\frac{4}{17}$ than to list the multiples of both 9 and 17.

Students will also compare fractions to benchmarks of $\frac{1}{2}$ and 1 when a common numerator or denominator cannot be easily found.

Improper Fractions

Fractions less than 1 are called proper fractions, while fractions equal to or greater than 1 are called improper fractions. "Improper" does not mean there is anything wrong with these fractions. It is acceptable to express answers as improper fractions.

It is easier to see the size of the number when we express improper fractions as mixed numbers. For example, $2\frac{1}{2}$ cups flour is easier to visualize than $\frac{5}{2}$ cups flour.

Students will be asked to express all their final answers in simplest form, and mixed numbers are considered for this purpose to be simplest form.

Students should understand that the numerator counts fractional units and that the fractional units can be counted indefinitely. These counts continue forever, so improper fractions continue beyond 1. Additionally, it is important to have students see that large numbers can be expressed as improper fractions.

Students will relate improper fractions to mixed numbers on a number line:

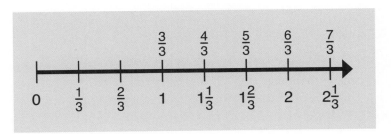

Since $1 = \frac{3}{3}$, $\frac{7}{3}$ can be thought of as $\frac{3}{3} + \frac{3}{3} + \frac{1}{3}$, or $2\frac{1}{3}$.

Chapter 6 Fractions

Notes & Materials

Mixed Numbers

A mixed number is the sum of a whole number and a proper fraction: 1 and $\frac{3}{5}$ is $1 + \frac{3}{5}$.

$1\frac{3}{5}$ has units of ones (1) and units of fifths (3).

To avoid confusion when writing fractions, the horizontal line notation should be used instead of a slanted line. Students using a slanted line may confuse a mixed number with an improper fraction. For example:

$$1\frac{3}{5} \qquad 13/5$$

Is this $1\frac{3}{5}$ or 13/5?

Adding and Subtracting Mixed Numbers

To develop their understanding of how mixed numbers are expressed, students will add fractions to and subtract fractions from a whole number. Addition is straightforward:

$3 + \frac{5}{8} = 3\frac{5}{8}$

Students learned to subtract a fraction from 1 in Dimensions Math 3. To subtract a fraction from a whole number greater than 1, they can mentally split the number into 1 and the remaining amount, subtract the fraction from 1, and add it to the difference.

$$3 - \frac{5}{8} = 2 + \frac{3}{8}$$
$$\diagup \quad \diagdown$$
$$2 \qquad \frac{8}{8}$$

Students will convert between improper fractions and mixed numbers by grouping fractions to make whole numbers:

$2\frac{1}{3} = 2 + \frac{1}{3}$
$= \frac{6}{3} + \frac{1}{3} = \frac{7}{3}$

Converting a mixed number to an improper fraction involves finding out how many unit fractions are in the mixed number. Converting an improper fraction to a mixed number involves finding how many whole numbers can be created from the improper fraction.

In Lesson 7, students will formally learn about the relationship between division and fractions, and how to express the remainder in a division problem as a fraction.

Materials

- Dry erase markers
- Fraction bars or fifths
- Index cards
- Pattern blocks
- Ruler or straightedge
- String
- Whiteboards

Blackline Masters

- Equivalent Fractions
- Equivalent Fractions Number Lines
- Human Number Line Cards
- Improper Fractions to Mixed Fractions Puzzle
- Mixed Number Match Cards
- Number Cards
- Order Up Improper Fractions and Mixed Number Cards

Activities

Activities included in this chapter provide practice with fractions and mixed numbers. They can be used after students complete the **Do** questions, or anytime additional practice is needed.

Chapter Opener

Objective

- Review fraction concepts.

Discuss the examples in the **Chapter Opener**. Ensure students know the terms "numerator" and "denominator". Check to see if students recall equivalent fractions.

The circle has $\frac{6}{8}$ of its fractional parts shaded.

The triangle has $\frac{3}{4}$ of its fractional parts shaded, which is equivalent to $\frac{6}{8}$.

The colored part of the hexagon is $\frac{18}{24}$ or $\frac{6}{8}$.

The colored part of the rectangle shows $\frac{9}{12}$. Students can simplify $\frac{9}{12}$ to $\frac{3}{4}$ to see that $\frac{9}{12}$ is also equal to $\frac{6}{8}$.

Activity

▲ **Pattern Block Fractions**

Materials: Pattern blocks

Provide students with pattern blocks. Name one piece as 1 whole and have them determine the values of the other pieces in the set.

For example:

- If the hexagon is 1 whole, what fraction of the whole is the green triangle? The blue rhombus? The red trapezoid?
- If the trapezoid is 1 whole, what fraction of the whole is the green triangle?

If ⬡ = 1 then ▲ = $\frac{1}{6}$.

Have students make different fractions with the pattern blocks.

Chapter 6

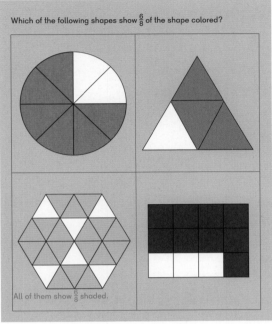

Fractions

Which of the following shapes show $\frac{6}{8}$ of the shape colored?

All of them show $\frac{6}{8}$ shaded.

To help students recall equivalent fractions, ask questions such as:

- "If the hexagon is 1 whole, what fraction of the whole is the trapezoid?"
- "Can you use other fractional parts (or shapes) to create a trapezoid?"
- "Can you make a statement regarding equivalent fractions?"

Lesson 1 Equivalent Fractions

Objectives

- Review equivalent fractions.
- Use multiplication and division to generate equivalent fractions.

Lesson Materials

- Equivalent Fractions (BLM)
- Equivalent Fractions Number Lines (BLM)
- Ruler or straightedge

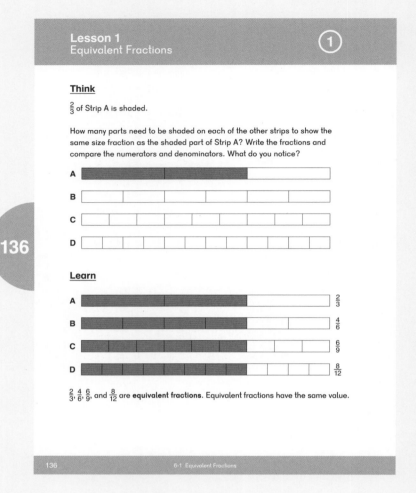

Think

Provide students with Equivalent Fractions (BLM) and pose the **Think** problem. Students will shade the boxes and write each fraction of the colored part of the whole strip.

Have students discuss what they notice about the equivalent fractions.

Students may notice:

- When a whole is cut into more equal parts to find equivalent fractions, more parts are shaded.
- To find an equivalent fraction, the numerator and denominator can be multiplied or divided by the same number.
- The numerators and denominator are the same multiples of 2 and 3, similar to the example in **Learn**.

Learn

Students can compare their equivalent fractions to the example in **Learn**.

Have students look at their colored Equivalent Fractions (BLM) and read Dion's comment. Multiplying and dividing numerators and denominators to find an equivalent fraction does not actually change the size of the shaded part of the strips. If we double the number of parts, the size of each part will be half as large but the shaded area will remain the same.

Have students find another fraction equivalent to $\frac{2}{3}$.

There is no whole number we can multiply the numerator and denominator of $\frac{6}{9}$ by to show it is equivalent to $\frac{8}{12}$. However, both of these fractions are equivalent to $\frac{2}{3}$.

Emma points out that we can show $\frac{8}{12}$ and $\frac{6}{9}$ are equivalent because they can each be simplified to $\frac{2}{3}$.

Review the term "simplest form" with students. Students can see that when we simplify a fraction, we are dividing the numerator and denominator by a common factor of both numbers.

Do

1 This question is included as Equivalent Fractions Number Lines (BLM) for students to complete and keep as a reference.

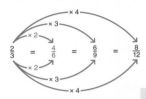

We can find equivalent fractions by multiplying or dividing the numerator and denominator by the same number.

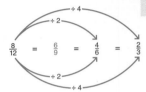

When we divide both the numerator and denominator by a common factor, we are **simplifying** the fraction.

The **simplest form** of $\frac{8}{12}$ is $\frac{2}{3}$ because 1 is the only common factor of both the numerator and denominator of $\frac{2}{3}$.

Do

1 Find equivalent fractions using these number lines.

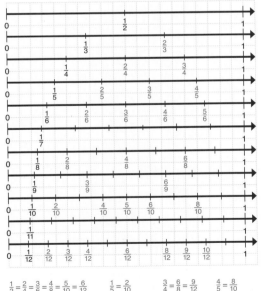

$\frac{1}{2} = \frac{2}{4} = \frac{3}{6} = \frac{4}{8} = \frac{5}{10} = \frac{6}{12}$ $\frac{1}{5} = \frac{2}{10}$ $\frac{3}{4} = \frac{6}{8} = \frac{9}{12}$ $\frac{4}{5} = \frac{8}{10}$

$\frac{1}{3} = \frac{2}{6} = \frac{3}{9} = \frac{4}{12}$ $\frac{1}{6} = \frac{2}{12}$ $\frac{2}{5} = \frac{4}{10}$ $\frac{5}{6} = \frac{10}{12}$

$\frac{1}{4} = \frac{2}{8} = \frac{3}{12}$ $\frac{2}{3} = \frac{4}{6} = \frac{6}{9} = \frac{8}{12}$ $\frac{3}{5} = \frac{6}{10}$

⑤ Students should notice that they can simplify a fraction in steps. In this problem, they see that they can divide by 2 twice, or by 4. Simplifying in steps is helpful with greater numbers as students can find any common factor to simply the fraction. If students use 4 as their factor, they will find simplest form in one step.

Exercise 1 • page 127

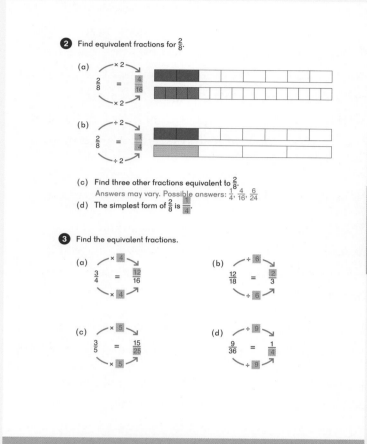

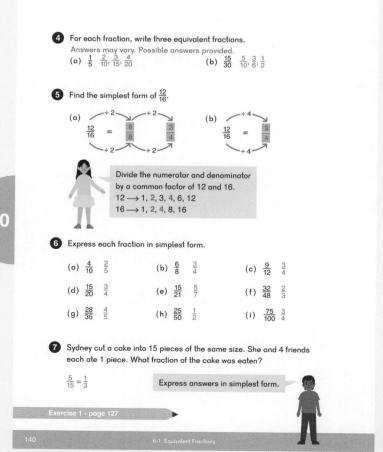

Lesson 2 Comparing and Ordering Fractions

Objective

- Compare fractions by finding common denominators or numerators.

Think

Pose the **Think** problem. Students should recall that there are several ways to compare the fractions $\frac{2}{3}$ and $\frac{4}{5}$.

Provide students with adequate time to work through the problem and challenge them to solve it in more than one way.

Have students share and discuss their solutions.

Learn

Have students compare their solutions from **Think** to the ones given in **Learn**.

Discuss Emma's and Dion's comments.

Emma thinks of a common multiple of 3 and 5. With a common denominator, she can compare the fractions by comparing the numerators because the fractional units (size of the parts) are the same.

Students often rely on only this strategy. Encourage them to try the two methods on the next page.

Sofia finds a common numerator. If the numerators are the same, the number of parts is the same. We can look at the size of each part (the denominator) to compare.

Mei points out that to compare fractions with like numerators you only need to compare the unit fractions.

Alex compares the fractions to 1. When fractions are both just a unit fraction from the whole, it is easy to compare them. $\frac{2}{3}$ is $\frac{1}{3}$ from the whole, and $\frac{4}{5}$ is $\frac{1}{5}$ from the whole.

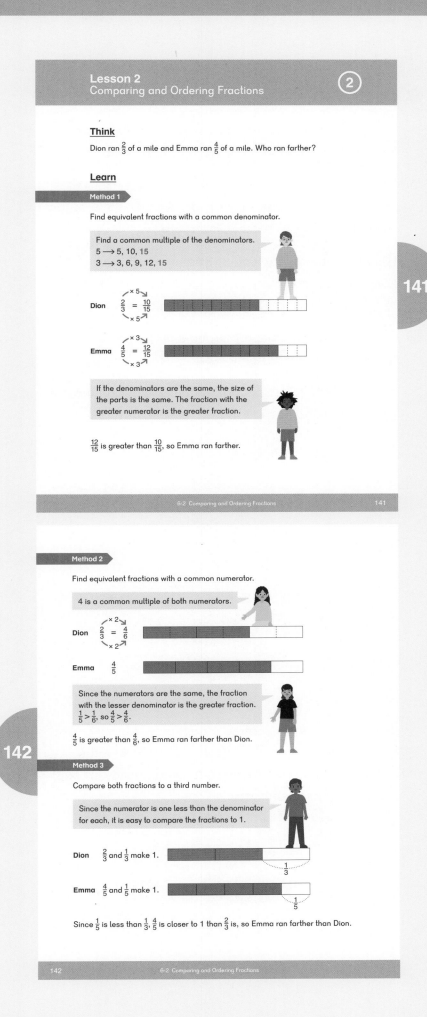

© 2019 Singapore Math Inc.   Teacher's Guide 4A Chapter 6   187

Do

1 (a) Sofia sees the same denominator.

(b) Dion compares the unit fractions. He knows $\frac{1}{8}$ is less than $\frac{1}{5}$, so three $\frac{1}{5}$s is greater than three $\frac{1}{8}$s.

(c) Alex finds an equivalent fraction for $\frac{3}{4}$. He notices that one numerator is an easy multiple of the other. He compares the numerators.

(d) Mei finds an equivalent fraction for $\frac{3}{5}$. She notices that one denominator is an easy multiple of the other. She compares the denominators.

2 Discuss how Emma's and Sofia's methods both found a common denominator.

Sofia knows that a common denominator can always be found by multiplying the two denominators.

$$\frac{3}{4} \xrightarrow{\times 10} \frac{30}{40}$$

$$\frac{7}{10} \xrightarrow{\times 4} \frac{28}{40}$$

By listing multiples, Emma also finds that any common multiple of the two denominators works. Multiplying denominators together will always result in a common denominator.

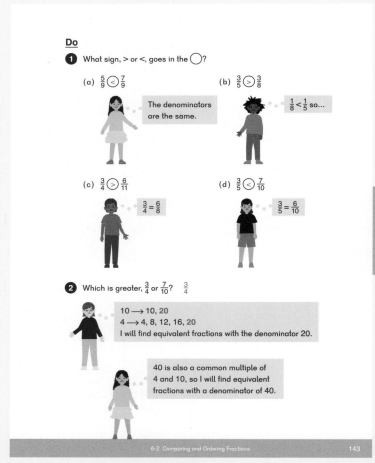

3. Dion compares both fractions to $\frac{1}{2}$ or $\frac{4}{8}$.

$\frac{3}{8}$ is less than $\frac{1}{2}$ ($\frac{4}{8}$).

$\frac{4}{5}$ is greater than $\frac{1}{2}$ ($\frac{4}{8}$).

So, $\frac{3}{8}$ is less than $\frac{4}{5}$.

5. Students should use a variety of strategies from the lesson when ordering the fractions.

Activity

▲ How Many?

Materials: Number Cards (BLM) 1–9

Using any of the digits 1–9 once, have students find fractions less than $\frac{1}{2}$.

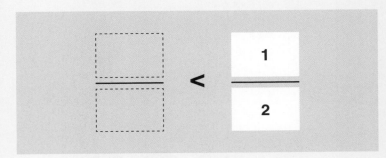

★ To extend the activity, have students find two fractions that are greater than $\frac{1}{4}$ and less than $\frac{1}{2}$.

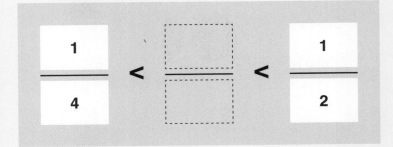

Exercise 2 • page 130

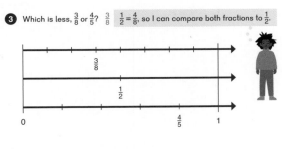

Lesson 3 Improper Fractions and Mixed Numbers

Objective

- Express fractions that are greater than 1 as improper fractions and mixed numbers.

Think

Pose the **Think** problem. Have students draw the number line and label each tick mark.

Discuss student solutions.

Learn

Students should recall that fractions continue on a number line beyond 1. Each tick mark represents one third, so they can count the tick marks in thirds: $\frac{1}{3}$, $\frac{2}{3}$, $\frac{3}{3}$, $\frac{4}{3}$, $\frac{5}{3}$.

Have students compare the numbers above and below the number line.

Mei introduces how to read a mixed number.

Introduce the terms in **Learn**: "proper fraction," "improper fraction," and "mixed number." Have students give other examples of each.

Alex tells students that we can think of a mixed number as the sum of a whole number and a fraction less than 1.

Do

② Students should see that since $\frac{2}{2} = 1$, $\frac{3}{2} = \frac{2}{2} + \frac{1}{2}$.

③ (a) Ensure students understand that the first 1 L beaker is full and 1 L = $\frac{5}{5}$ L. The second 1 L beaker is $\frac{3}{5}$ full so there are $\frac{3}{5}$ L. 5 fifths + 3 fifths = 8 fifths. Altogether there is a total of $\frac{8}{5}$, or $1\frac{3}{5}$ liters of water in the beakers.

(b) Dion prompts students to simplify their answers.

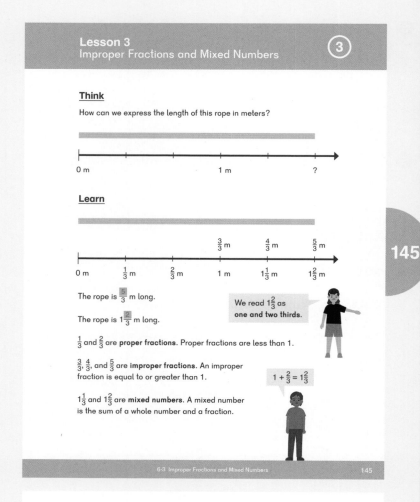

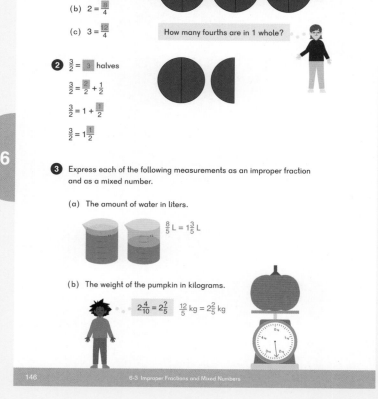

190   Teacher's Guide 4A Chapter 6   © 2019 Singapore Math Inc.

5 (b) Students can think about subtracting from 1 whole. They will convert between mixed numbers and improper fractions in Lesson 6.

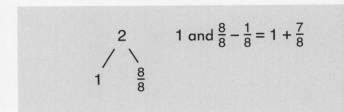

1 and $\frac{8}{8} - \frac{1}{8} = 1 + \frac{7}{8}$

(d)

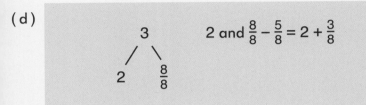

2 and $\frac{8}{8} - \frac{5}{8} = 2 + \frac{3}{8}$

Activity

▲ Mixed Bag

Materials: Pattern blocks — hexagons, triangles, trapezoids, rhombuses

The hexagon is 1 whole. Have students grab a handful of hexagons, triangles, trapezoids, or rhombuses. Have them express how many hexagons they have as a mixed number.

Example:

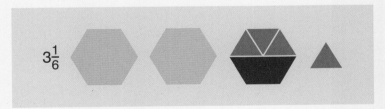

Exercise 3 • page 133

Lesson 4 Practice A

Objective

- Practice fraction concepts from the chapter.

After students complete the **Practice** in the textbook, have them continue to practice ordering and comparing fractions with activities from the chapter.

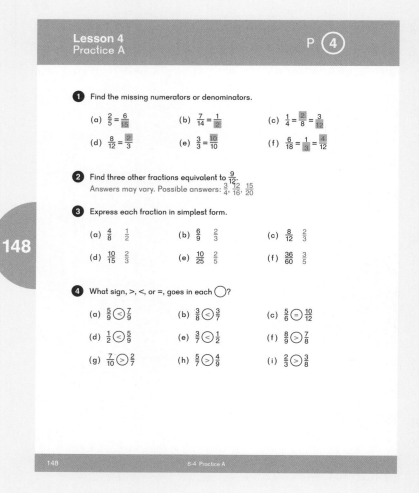

5 Not all of the fractions have the same denominator or numerator. Students can compare two fractions at a time.

8 Students may solve in multiple ways:

- Compare common numerators: $\frac{2}{3} = \frac{4}{6}$.
- Compare common denominators: $\frac{2}{3} = \frac{14}{21}$ and $\frac{4}{7} = \frac{12}{21}$.

Activity

▲ Number Line

Materials: Order Up Improper Fractions and Mixed Number Cards (BLM), string

Suspend the string or have students hold it up at either end. Begin by handing out the 0 and 1 cards to students and have them place the cards on the string number line.

Hand out cards one at a time and have students adjust the other cards in relation to the new cards, hanging the new cards in their appropriate places.

For example:

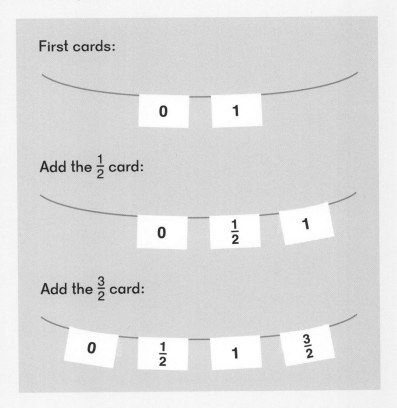

Note that not all cards need to be used for the activity.

Have students explain their reasoning when placing their cards on the number line. Equivalent fractions can be stacked on top of each other.

If students need additional help, have them discuss where each new card should be placed.

◀ **Exercise 4 • page 136**

Lesson 5 Expressing an Improper Fraction as a Mixed Number

Objective

- Express improper fractions as mixed numbers.

Lesson Materials

- Fraction bars for fifths

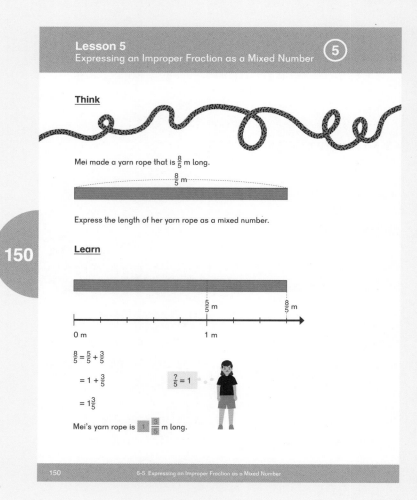

Think

Provide students with fraction bars and pose the **Think** problem. Students should see that the $\frac{8}{5}$ m long rope is also considered $1\frac{3}{5}$ m long.

Learn

Discuss the number line and what improper fractions or mixed numbers each tick mark represents.

Improper fractions can be converted into mixed numbers by thinking about the parts that make 1. Mei reminds us that $\frac{5}{5}$ is equal to 1. We can think of $\frac{8}{5}$ as $\frac{5}{5}$ and $\frac{3}{5}$.

Students should see that to change an improper fraction to a mixed number, we need to find the number of wholes first and then add the fractional part that is left.

Do

1 – **2** Discuss the problems with students.

1 (b) 2 wholes = $\frac{10}{5}$

2 Students can use the number line to help them find the missing mixed numbers.

3 Students should think of the number of wholes that can be made from the fractional units. For example, in (a), since $\frac{8}{2} = 4$, $\frac{9}{2} = 4\frac{1}{2}$.

4 Convert the distance Wyatt walked on Monday to a mixed number to compare.

Activity

▲ Mixed Number Memory

Materials: Mixed Number Match Cards (BLM)

Shuffle the cards and lay them facedown in an array. On each turn, players turn two cards faceup. If the cards are equivalent, the player keeps the cards and plays again.

If the cards are not equivalent, they are turned facedown and the player's turn is over.

The player with the most cards at the end of the game is the winner.

$1\frac{1}{2}$ $\frac{15}{10}$

Exercise 5 • page 139

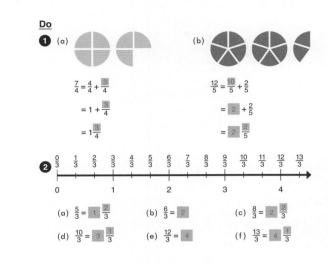

Lesson 6 Expressing a Mixed Number as an Improper Fraction

Objective

- Convert mixed numbers to improper fractions.

Lesson Materials

- Index cards, 3 for each student

Think

Provide students with three index cards and pose the **Think** problem. Tell them that one index card represents 1 whole.

Ask students to represent the amount of Alex's chocolate ($2\frac{3}{4}$ bars of chocolate) with the index cards.

Ask students to think about the first question (a). How many $\frac{1}{4}$ bar pieces will he have?

Finally, ask students to think about how they can express $2\frac{3}{4}$ bars of chocolate as an improper fraction.

Have them first draw or cut the index cards into fourths to make $2\frac{3}{4}$.

Students can either cross out or remove one of the fourths to ensure they start with $2\frac{3}{4}$ pieces of chocolate.

Discuss student solutions.

Learn

Mixed numbers can be converted into improper fractions by converting the whole number into an improper fraction, then adding the remaining fractional part.

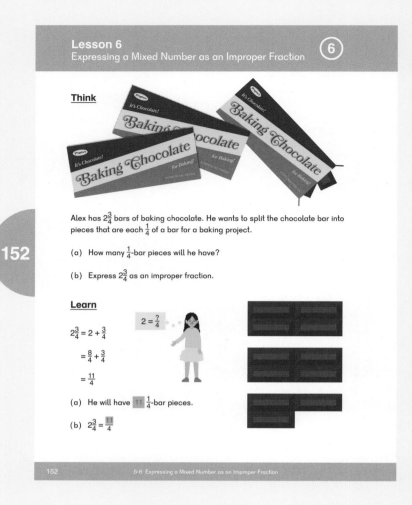

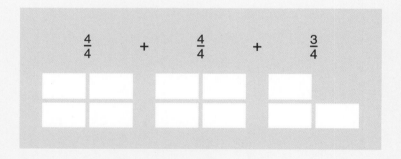

Sofia prompts students to find how many fourths are in 2.

Now that 2 is expressed in the same fractional unit, $\frac{8}{4}$, we add the fractional part $\frac{3}{4}$, to get $\frac{11}{4}$.

Do

② Students can use the number line to help them find the missing numerators.

④ (a) Students should be able to see without converting the numbers that $\frac{3}{4}$ is less than 1, $\frac{4}{3}$ is greater than 1 but less than 2, and $2\frac{1}{2}$ is greater than 2.

(b), (c) If students convert the mixed numbers to improper fractions, they will easily see the order of the fractions.

Activity

▲ **Human Number Line**

Materials: Human Number Line Cards (BLM)

Students each get one card and arrange themselves in order (based off of what is on their cards) from least to greatest. Students who have cards with equivalent improper fractions and mixed numbers stand in front of each other to show they mark the same point on the number line.

To make it more fun, play music and have students arrange themselves by the time the music stops.

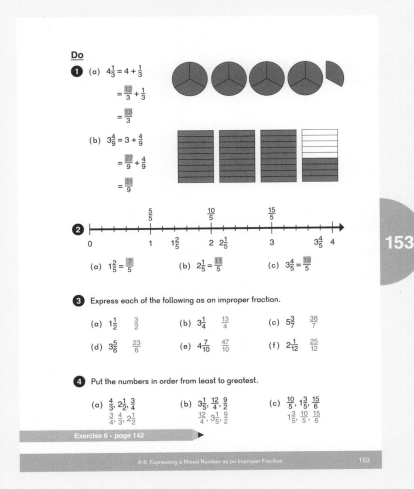

Lesson 7 Fractions and Division

Objective
- Understand fractions as division.

Lesson Materials
- Index cards, at least 7 for each student or pair of students

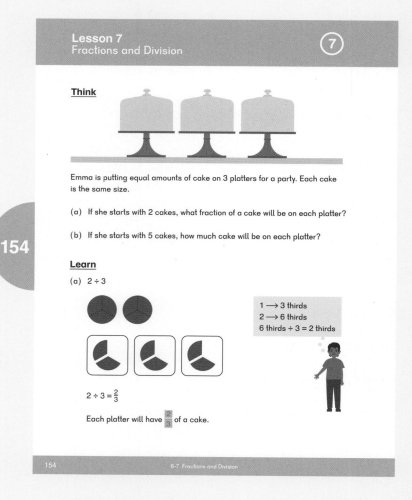

Think

Provide students with index cards to represent the cakes. Tell them that one index card represents 1 cake. Discuss the **Think** problem and have students draw or cut the cards to show solutions to questions (a) and (b).

Point out to students that for (a) they evenly divided 2 cakes among 3 platters. Ask what equation they could write to represent what they did with the index cards.

Have them think about an equation for question (b).

Learn

(a) Alex thinks that he can first divide 1 cake into 3 equal pieces (1 for each platter). Then he can do the same with the next cake. That is, if 1 whole is 3 thirds, then 2 wholes must be 2 × 3 thirds or 6 thirds. If we think of thirds as units then 6 thirds ÷ 3 = 2 thirds.

Students should see that the line between the numerator and denominator in a fraction means division.

Students may also have divided each card into halves and divided the halves among 3 platters. Since they have 1 half left to divide, they can cut that half into three equal pieces.

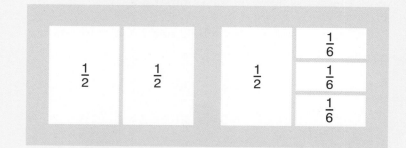

Each platter has $\frac{1}{2} + \frac{1}{6}$ of a cake, or $\frac{3}{6} + \frac{1}{6} = \frac{4}{6} = \frac{2}{3}$ of a cake.

(b) To divide 5 by 3, Emma first divides each of her 5 wholes into thirds. She then divides the 15 thirds among the 3 platters. Each platter will have 5 one-third pieces, so each platter will have $\frac{5}{3}$ of a cake.

Dion puts 3 whole cakes on the 3 platters first. Each platter has 1 cake. He has 2 whole cakes left. Similar to what Alex does in (a), Dion divides each of the remaining cakes into 3 equal pieces, or thirds. Each platter will have $1\frac{2}{3}$ cake.

Dion relates this to the division algorithm. 5 divided by 3 is $1\frac{2}{3}$, because the remainder, 2, is being divided by 3.

Do

1. In each of these, a length of ribbon is divided into 4 equal pieces.

In (a), the initial piece is 1 whole, which is 1 m, so each equal part is $\frac{1}{4}$. In (b) the whole is greater than 1, but each part is less than 1 and therefore still a fraction. In (c), each part is greater than 1, but not a whole number, and can be expressed as a fraction.

In each of these, the answer to the division problem is written as a fraction or a mixed number.

(c) Mei points out that we can continue dividing the remainder and find a fractional answer. This method can be used to convert an improper fraction to a mixed number.

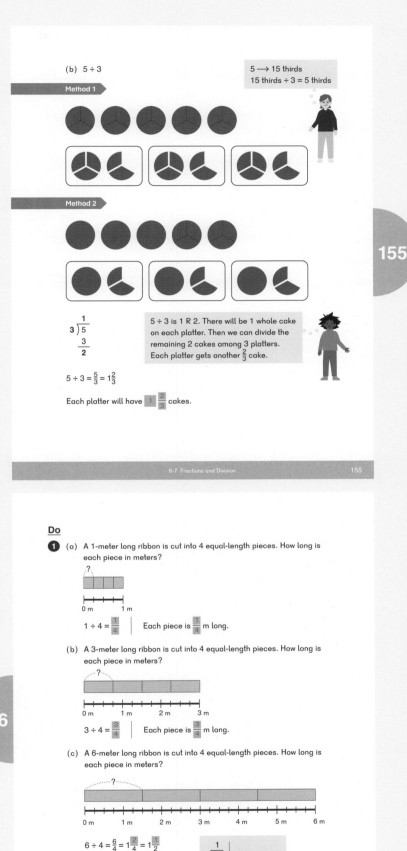

2 (a) Sofia knows that 3 liters of water divided equally among 4 bottles will result in less than 1 liter of water in each bottle. If students need help to see this, have them work with 1 liter at a time.

Repeat for each liter.

(b) Dion sees that there are more whole liters of water than there are bottles, so each bottle will get at least 1 liter.

3 Emma converts an improper fraction to a mixed number the way students have previously learned. Alex shows that students can now use a different method: division.

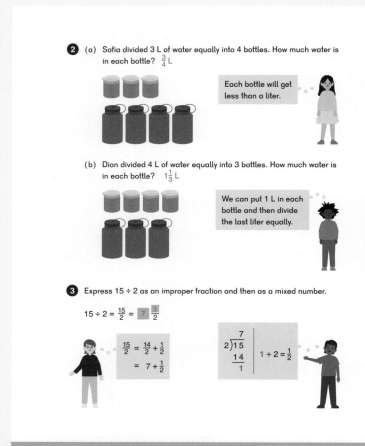

④ To express $2\frac{5}{8}$ as an improper fraction, Alex thinks 8 eighths and 8 eighths or 2 × 8 eighths. He adds the $\frac{5}{8}$ to $\frac{16}{8}$.

When you multiply the denominator by the whole number you get the number of eighths in 2:

$1 = \frac{8}{8}$, so $2 = \frac{16}{8}$.

⑤ Sofia uses Alex's method. She knows that 5 fifths + 5 fifths + 5 fifths is the same as 3 × 5 fifths. She adds $\frac{4}{5}$ to $\frac{15}{5}$.

$1 = \frac{5}{5}$

$2 = \frac{10}{5}$

$3 = \frac{15}{5}$

Multiply the whole number, 3, by the denominator to find the number of fifths in 3.

Activity

★ How Many?

Materials: Number Cards (BLM) 0–9

Using six of the digits 0–9 once, have students find as many equivalent fractions as possible.

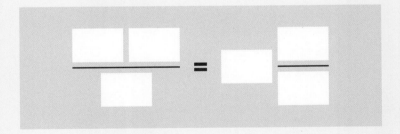

Sample answers:

$\frac{38}{4} = 9\frac{1}{2}$

$\frac{50}{6} = 8\frac{1}{3}$

$\frac{62}{8} = 7\frac{3}{4}$

Lesson 8 Practice B

Objective

- Practice concepts from the chapter.

After students complete the **Practice** in the textbook, have them continue to practice ordering and comparing fractions with activities from the chapter.

Exercise 8 • page 148

Brain Works

★ Equivalent Fractions Puzzle

Materials: Improper Fractions to Mixed Fractions Puzzle (BLM)

To prepare, print Improper Fractions to Mixed Fractions Puzzle (BLM) on cardstock for each student or pair of students and cut out triangles along the bold lines.

Students match the sides of the triangles with an equivalent fraction.

The puzzle will form a hexagon when complete.

Solution:

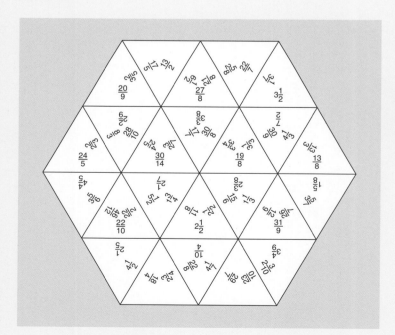

Chapter 6 Fractions

Exercise 1

Basics

1 Write the equivalent fractions.

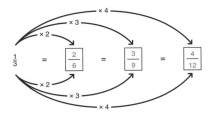

2 Simplify the fractions.

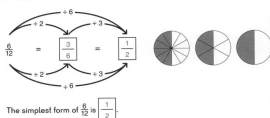

The simplest form of $\frac{6}{12}$ is $\frac{1}{2}$.

Practice

3 Write the missing numbers for the fraction indicated by the arrow.

$\frac{3}{4} = \frac{6}{8} = \frac{9}{12} = \frac{12}{16}$

4 Label the tick marks indicated by the arrows as fractions in simplest form.

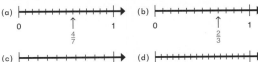

(a) $\frac{4}{7}$ (b) $\frac{2}{3}$ (c) $\frac{4}{5}$ (d) $\frac{3}{8}$

5 Express each fraction in simplest form.

(a) $\frac{6}{9} = \frac{2}{3}$ (b) $\frac{6}{8} = \frac{3}{4}$

(c) $\frac{6}{15} = \frac{2}{5}$ (d) $\frac{10}{12} = \frac{5}{6}$

(e) $\frac{21}{42} = \frac{1}{2}$ (f) $\frac{24}{32} = \frac{3}{4}$

(g) $\frac{42}{56} = \frac{3}{4}$ (h) $\frac{18}{72} = \frac{1}{4}$

6 Write the missing numbers.

(a) $\frac{28}{42} = \frac{14}{21} = \frac{2}{3} = \frac{6}{9} = \frac{18}{27}$

(b) $\frac{75}{100} = \frac{15}{20} = \frac{3}{4} = \frac{9}{12} = \frac{27}{36}$

7 Find the equivalent fractions.

(a) $\frac{5}{10} = \frac{1}{2}$ (b) $\frac{1}{2} = \frac{8}{16}$ (c) $\frac{5}{10} = \frac{8}{16}$

(d) $\frac{6}{9} = \frac{2}{3}$ (e) $\frac{2}{3} = \frac{8}{12}$ (f) $\frac{6}{9} = \frac{8}{12}$

(g) $\frac{6}{8} = \frac{3}{4}$ (h) $\frac{3}{4} = \frac{9}{12}$ (i) $\frac{6}{8} = \frac{9}{12}$

(j) $\frac{8}{10} = \frac{4}{5}$ (k) $\frac{12}{15} = \frac{4}{5}$ (l) $\frac{8}{10} = \frac{12}{15}$

(m) $\frac{3}{6} = \frac{4}{8}$ (n) $\frac{12}{16} = \frac{15}{20}$ (o) $\frac{24}{30} = \frac{16}{20}$

Challenge

8 The fractions below form a number pattern. What is the missing fraction?

$\frac{5}{12}, \frac{1}{2}, \frac{7}{12}, \frac{2}{3}, \frac{3}{4}$

The numbers increase by $\frac{1}{12}$.

Exercise 2 • pages 130–132

Exercise 2

Basics

1 Write > or < in each ◯.

(a) $\frac{5}{12}$ ⊘ $\frac{7}{12}$

(b) $\frac{5}{12}$ ⊘ $\frac{5}{9}$

2 Label the correct tick marks for the given fractions on the number lines. Then find equivalent fractions. Write > or < in each ◯.

$\frac{2}{3} = \boxed{\frac{8}{12}}$ $\frac{7}{12}$ $\frac{2}{3}$ ⊘ $\frac{7}{12}$

$\frac{3}{5} = \boxed{\frac{9}{15}}$ $\frac{9}{16}$ $\frac{3}{5}$ ⊘ $\frac{9}{16}$

$\frac{2}{5} = \boxed{\frac{14}{35}}$ or $\frac{6}{15}$ $\frac{3}{7} = \boxed{\frac{15}{35}}$ or $\frac{6}{14}$ $\frac{2}{5}$ ⊘ $\frac{3}{7}$

---

3 Write >, <, or = in each ◯.

(a) $\frac{4}{8}$ ⊘ $\frac{1}{2}$ $\frac{5}{8}$ ⊘ $\frac{1}{2}$ $\frac{3}{8}$ ⊘ $\frac{1}{2}$

(b) $\frac{6}{12}$ ⊘ $\frac{1}{2}$ $\frac{6}{13}$ ⊘ $\frac{1}{2}$ $\frac{6}{11}$ ⊘ $\frac{1}{2}$

(c) $\frac{6}{13}$ ⊘ $\frac{5}{8}$ $\frac{6}{11}$ ⊘ $\frac{3}{8}$

4 (a) Label $\frac{4}{5}$ and $\frac{6}{7}$ on the number line.

(b) $\frac{4}{5}$ and $\boxed{\frac{1}{5}}$ make 1. $\frac{6}{7}$ and $\boxed{\frac{1}{7}}$ make 1.

(c) Which fraction is closer to 1?
$\frac{6}{7}$
Which fraction is greater?
$\frac{6}{7}$

Practice

5 Write > or < in each ◯.

(a) $\frac{7}{8}$ ⊘ $\frac{3}{4}$ (b) $\frac{3}{5}$ ⊘ $\frac{7}{15}$

(c) $\frac{4}{7}$ ⊘ $\frac{8}{9}$ (d) $\frac{3}{4}$ ⊘ $\frac{9}{11}$

(e) $\frac{2}{3}$ ⊘ $\frac{3}{8}$ (f) $\frac{5}{12}$ ⊘ $\frac{7}{16}$

(g) $\frac{10}{11}$ ⊘ $\frac{8}{9}$ (h) $\frac{15}{16}$ ⊘ $\frac{10}{11}$

---

6 Circle the fractions that are less than $\frac{1}{2}$. Then write the fractions in order from least to greatest.

(⊘$\frac{3}{8}$) $\frac{8}{11}$ $\frac{3}{5}$ (⊘$\frac{5}{16}$) $\frac{8}{15}$ (⊘$\frac{5}{12}$)

$\frac{5}{16}, \frac{3}{8}, \frac{5}{12}, \frac{8}{15}, \frac{3}{5}, \frac{8}{11}$

7 Write the fractions in order from least to greatest.

(a) $\frac{5}{6}, \frac{6}{7}, \frac{3}{5}$ $\frac{3}{5}, \frac{5}{6}, \frac{6}{7}$

(b) $\frac{8}{9}, \frac{23}{50}, \frac{12}{13}$ $\frac{23}{50}, \frac{8}{9}, \frac{12}{13}$ Compare each to $\frac{1}{2}$ or 1 first, and then compare those less than $\frac{1}{2}$ and greater than $\frac{1}{2}$ to each other, finding equivalent fractions with either common numerators or common denominators.

(c) $\frac{13}{16}, \frac{2}{5}, \frac{5}{8}, \frac{7}{7}$ $\frac{2}{5}, \frac{5}{8}, \frac{13}{16}, \frac{7}{7}$

(d) $\frac{7}{12}, \frac{3}{7}, \frac{5}{14}, \frac{31}{48}$ $\frac{5}{14}, \frac{3}{7}, \frac{7}{12}, \frac{31}{48}$

Challenge

8 Use the given numbers to fill in the missing numerators or denominators so that the fractions are in order from least to greatest. Each fraction should be less than 1 and in simplest form.

(a) 5, 6, 7, 8 $\boxed{\frac{3}{8}} < \boxed{\frac{7}{12}} < \boxed{\frac{5}{6}}$

(b) 7, 8, 9, 10, 11, 12 $\boxed{\frac{7}{8}} < \boxed{\frac{9}{10}} < \boxed{\frac{11}{12}}$

Answers may vary.

---

204   Teacher's Guide 4A Chapter 6   © 2019 Singapore Math Inc.

Exercise 3 • pages 133–135

Exercise 3

Basics

1. Write an improper fraction for each of the following.

   (a) 8 fifths = $\dfrac{8}{5}$

   (b) 9 thirds = $\dfrac{9}{3}$

   (c) 11 sevenths = $\dfrac{11}{7}$

2. Write a mixed number for each of the following.

   (a) $2 + \dfrac{5}{9} = 2\dfrac{5}{9}$

   (b) $1 + \dfrac{8}{12} = 1\dfrac{2}{3}$

   (c) $3 + \dfrac{2}{3} = 3\dfrac{2}{3}$

Practice

3. Express each of the following as a mixed number and as an improper fraction in simplest form.

	Mixed Number	Improper Fraction
	$2\dfrac{1}{8}$	$\dfrac{17}{8}$
	$3\dfrac{5}{6}$	$\dfrac{23}{6}$
	$4\dfrac{4}{9}$	$\dfrac{40}{9}$

4. Finish labeling each arrow with a fraction above the number line and a mixed number below the number line. Use simplest form.

   Above: $\dfrac{7}{12}$, $\dfrac{5}{6}$, $\dfrac{3}{2}$, $\dfrac{7}{4}$, $\dfrac{29}{12}$, $\dfrac{19}{6}$

   Below: $1\dfrac{1}{2}$, $1\dfrac{3}{4}$, $2\dfrac{5}{12}$, $3\dfrac{1}{6}$

5. Write the length of each line in centimeters as both a mixed number and a fraction. Use simplest form.

   A $10\dfrac{3}{10}$, $\dfrac{103}{10}$

   B $7\dfrac{4}{5}$, $\dfrac{39}{5}$

6. Write a mixed number for each of the following.

   (a) $3 - \dfrac{1}{5} = 2\dfrac{4}{5}$

   (b) $4 - \dfrac{2}{9} = 3\dfrac{7}{9}$

7. Express each value as a mixed number in simplest form.

   (a) $3 + \dfrac{2}{3}$ $\quad 3\dfrac{2}{3}$

   (b) $4 - \dfrac{4}{9}$ $\quad 3\dfrac{5}{9}$

   (c) $5 + \dfrac{6}{8}$ $\quad 5\dfrac{3}{4}$

   (d) $8 - \dfrac{6}{12}$ $\quad 7\dfrac{1}{2}$

   (e) $\dfrac{8}{14} + 8$ $\quad 8\dfrac{4}{7}$

   (f) $3 - \dfrac{7}{15}$ $\quad 2\dfrac{8}{15}$

Exercise 4 • pages 136–138

Exercise 4

Check

1. Write the fraction for the shaded part of each shape in simplest form.

   (a)  $\boxed{\dfrac{3}{7}}$ (b) $\boxed{\dfrac{2}{5}}$

2. Which fractions are greater than 1?

   $\left(\dfrac{8}{5}\right)$ $\dfrac{7}{16}$ $\left(\dfrac{11}{9}\right)$ $\left(\dfrac{15}{13}\right)$ $\dfrac{6}{15}$

3. Which fraction is closest to 1?

   $\dfrac{3}{4}$ $\dfrac{1}{2}$ $\dfrac{4}{5}$ $\left(\dfrac{5}{6}\right)$ $\dfrac{2}{3}$

4. Write > or < in each ◯.

   (a) $\dfrac{5}{8}\ \boxed{<}\ \dfrac{8}{5}$ (b) $2\dfrac{3}{5}\ \boxed{>}\ 2\dfrac{7}{15}$ (c) $\dfrac{11}{8}\ \boxed{>}\ \dfrac{5}{4}$

5. In a relay race, Aisha ran $1\dfrac{3}{4}$ of a mile and Paula ran $1\dfrac{5}{12}$ of a mile. Who ran farther?

   $1\dfrac{3}{4} > 1\dfrac{5}{12}$

   Aisha

6. Label each arrow with a fraction above the number line and a mixed number below the number line. Use simplest form.

   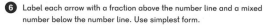
   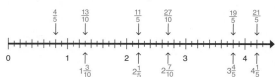

7. Label the tick marks on the number line with the given numbers.

   $3\dfrac{3}{10},\ 2\dfrac{2}{5},\ \dfrac{1}{2},\ 1\dfrac{1}{3},\ \dfrac{9}{12},\ 3\dfrac{7}{14},\ 1\dfrac{5}{7},\ 2\dfrac{6}{8}$

8. Write the missing numerators or denominators. Each fraction should be in simplest form.

   (a) $\dfrac{2}{9} < \boxed{\dfrac{1}{4}} < \dfrac{2}{7}$ (b) $\dfrac{1}{2} < \boxed{\dfrac{3}{5}} < \dfrac{3}{4}$

   (c) $\dfrac{3}{4} < \boxed{\dfrac{4}{5}} < 1$ (d) $2\dfrac{2}{3} < 2\boxed{\dfrac{8}{11}} < 2\dfrac{4}{5}$

Challenge

9. List all the different fractions between 0 and 1 that are in simplest form where the denominators are 10 or less.

   $\dfrac{1}{10}, \dfrac{1}{9}, \dfrac{1}{8}, \dfrac{1}{7}, \dfrac{1}{6}, \dfrac{1}{5}, \dfrac{1}{4}, \dfrac{1}{3}, \dfrac{1}{2}, \dfrac{2}{9}, \dfrac{2}{7}, \dfrac{2}{5}, \dfrac{2}{3}, \dfrac{3}{10}, \dfrac{3}{8}, \dfrac{3}{7}, \dfrac{3}{5}, \dfrac{3}{4}, \dfrac{4}{9}, \dfrac{4}{7}, \dfrac{4}{5}, \dfrac{5}{9}, \dfrac{5}{8}, \dfrac{5}{7}, \dfrac{5}{6}, \dfrac{6}{7}, \dfrac{7}{10}, \dfrac{7}{9}, \dfrac{7}{8}, \dfrac{8}{9}, \dfrac{9}{10}$

10. (a) How many more squares need to be shaded in order to have $\dfrac{4}{9}$ of the rectangle unshaded?

    5 more squares

    (b) How many more squares need to be shaded in order to have $\dfrac{1}{4}$ of the rectangle unshaded?

    12 more squares

    (c) How many more squares need to be shaded in order to have $1\dfrac{2}{3}$ unshaded?

    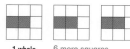

    1 whole   6 more squares

Teacher's Guide 4A Chapter 6 © 2019 Singapore Math Inc.

Exercise 5 • pages 139–141

Exercise 5

Basics

1 (a) Express $\frac{15}{4}$ as a mixed number.

$\boxed{\frac{4}{4}} = 1 \qquad \boxed{\frac{8}{4}} = 2 \qquad \boxed{\frac{12}{4}} = 3$

$\frac{15}{4} = \frac{12}{4} + \frac{3}{4}$

$= 3 + \frac{3}{4}$

$= 3\frac{3}{4}$

(b) Express $\frac{23}{9}$ as a mixed number.

$\boxed{\frac{9}{9}} = 1 \qquad \boxed{\frac{18}{9}} = 2$

$\frac{23}{9} = \boxed{\frac{18}{9}} + \boxed{\frac{5}{9}}$

$= 2 + \boxed{\frac{5}{9}}$

$= 2\frac{5}{9}$

Practice

2 Label each arrow with a mixed number in simplest form or a whole number.

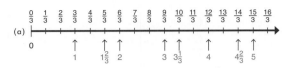
(a) arrows point to: 1, $1\frac{2}{3}$, 2, 3, $3\frac{1}{3}$, 4, $4\frac{2}{3}$, 5

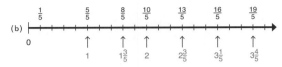
(b) arrows point to: 1, $1\frac{3}{5}$, 2, $2\frac{3}{5}$, $3\frac{1}{5}$, $3\frac{4}{5}$

(c) arrows point to: 1, $1\frac{3}{8}$, $1\frac{5}{8}$, $1\frac{7}{8}$, $2\frac{3}{8}$

3 Express each improper fraction as a whole number.

(a) $\frac{15}{5}$ 3   (b) $\frac{90}{9}$ 10

(c) $\frac{24}{2}$ 12   (d) $\frac{40}{8}$ 5

(e) $\frac{54}{6}$ 9   (f) $\frac{56}{7}$ 8

4 Express each improper fraction as a mixed number in simplest form.

(a) $\frac{18}{4} = \frac{16}{4} + \frac{2}{4}$ (b) $\frac{27}{5} = \frac{25}{5} + \frac{2}{5}$

    $= 4\frac{1}{2}$ $= 5\frac{2}{5}$

(c) $\frac{16}{7} = \frac{14}{7} + \frac{2}{7}$ (d) $\frac{9}{4} = \frac{8}{4} + \frac{1}{4}$

    $= 2\frac{2}{7}$ $= 2\frac{1}{4}$

(e) $\frac{20}{6} = \frac{18}{6} + \frac{2}{6}$ (f) $\frac{11}{2} = \frac{10}{2} + \frac{1}{2}$

    $= 3\frac{1}{3}$ $= 5\frac{1}{2}$

(g) $\frac{20}{8} = \frac{16}{8} + \frac{4}{8}$ (h) $\frac{30}{9} = \frac{27}{9} + \frac{3}{9}$

    $= 2\frac{1}{2}$ $= 3\frac{1}{3}$

5 Write the following numbers in order from least to greatest.

$\frac{18}{5}, \frac{23}{7}, \frac{14}{3}$

$\frac{23}{7}, \frac{18}{5}, \frac{14}{3}$

Exercise 6 • pages 142–144

Exercise 6

Basics

1 (a) Express $3\frac{2}{3}$ as an improper fraction.

$1 = \frac{3}{3}$ $2 = \frac{6}{3}$ $3 = \boxed{\frac{9}{3}}$

$3\frac{2}{3} = 3 + \frac{2}{3}$

$= \frac{9}{3} + \frac{2}{3}$

$= \frac{11}{3}$

(b) Express $2\frac{5}{8}$ as an improper fraction.

$1 = \boxed{\frac{8}{8}}$ $2 = \boxed{\frac{16}{8}}$

$2\frac{5}{8} = 2 + \boxed{\frac{5}{8}}$

$= \boxed{\frac{16}{8}} + \boxed{\frac{5}{8}}$

$= \frac{21}{8}$

Practice

2 Label each arrow with an improper fraction with a denominator of 4.

3 Label each arrow with an improper fraction.

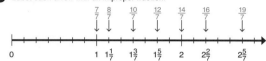

4 Express each whole number as an improper fraction with the given denominator.

(a) $2 = \boxed{\frac{18}{9}}$ (b) $3 = \boxed{\frac{24}{8}}$

(c) $4 = \boxed{\frac{20}{5}}$ (d) $5 = \boxed{\frac{15}{3}}$

(e) $4 = \boxed{\frac{48}{12}}$ (f) $8 = \boxed{\frac{128}{16}}$

5 Express each value as an improper fraction.

(a) $3\frac{2}{5} = \frac{15}{5} + \frac{2}{5}$ (b) $5\frac{2}{9} = \frac{45}{9} + \frac{2}{9}$

$= \frac{17}{5}$ $= \frac{47}{9}$

(c) $4\frac{5}{6} = \frac{24}{6} + \frac{5}{6}$ (d) $7\frac{1}{7} = \frac{49}{7} + \frac{1}{7}$

$= \frac{29}{6}$ $= \frac{50}{7}$

(e) $5\frac{7}{10} = \frac{50}{10} + \frac{7}{10}$ (f) $6\frac{5}{8} = \frac{48}{8} + \frac{5}{8}$

$= \frac{57}{10}$ $= \frac{53}{8}$

(g) $2\frac{5}{12} = \frac{24}{12} + \frac{5}{12}$ (h) $11\frac{2}{3} = \frac{33}{3} + \frac{2}{3}$

$= \frac{29}{12}$ $= \frac{35}{3}$

6 Write the following numbers in order from least to greatest.

$\frac{14}{3}, 3\frac{3}{7}, \frac{21}{4}$

$3\frac{3}{7}, \frac{14}{3}, \frac{21}{4}$

Exercise 7 • pages 145–147

Exercise 7

Basics

1 (a) Divide 2 by 5. Write the answer as a fraction.

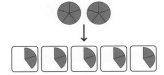

$2 \div 5 = \boxed{\dfrac{2}{5}}$

(b) Divide 7 by 5. Write the answer as a mixed number.

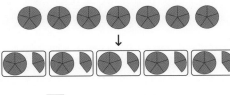

$7 \div 5 = \boxed{1\dfrac{2}{5}}$

$$5\overline{)7} \quad \boxed{1} \\ \boxed{5} \\ \boxed{2}$$

2 (a) Divide 2 by 4. Write the answer as a fraction in simplest form.

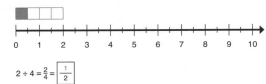

$2 \div 4 = \dfrac{2}{4} = \boxed{\dfrac{1}{2}}$

(b) Divide 10 by 4. Write the answer as a mixed number in simplest form.

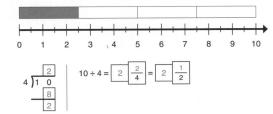

$4\overline{)10} \quad \boxed{2} \\ \boxed{8} \\ \boxed{2}$

$10 \div 4 = 2\boxed{\dfrac{2}{4}} = 2\boxed{\dfrac{1}{2}}$

3 Express each improper fraction as a mixed number in simplest form.

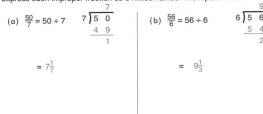

(a) $\dfrac{50}{7} = 50 \div 7$ $\quad 7\overline{)50}$, quotient 7 remainder 1

$= 7\dfrac{1}{7}$

(b) $\dfrac{56}{6} = 56 \div 6$ $\quad 6\overline{)56}$, quotient 9 remainder 2

$= 9\dfrac{1}{3}$

Practice

4 Divide. Express answers 1 or greater as whole or mixed numbers. Use simplest form.

(a) $3 \div 7$ $\dfrac{3}{7}$

(b) $8 \div 10$ $\dfrac{4}{5}$

(c) $\dfrac{45}{3}$ 15

(d) $31 \div 9$ $3\dfrac{4}{9}$

(e) $32 \div 6$ $5\dfrac{1}{3}$

(f) $49 \div 5$ $9\dfrac{4}{5}$

(g) $\dfrac{45}{4}$ $11\dfrac{1}{4}$

(h) $79 \div 5$ $15\dfrac{4}{5}$

5 A 38 oz box of cereal has 8 servings. How many ounces are in each serving?

$38 \div 8 = \dfrac{38}{8} = 4\dfrac{3}{4}$

$4\dfrac{3}{4}$ ounces

Exercise 8 • pages 148–150

Exercise 8

Check

1. Express each mixed number as an improper fraction.

   (a) $9\frac{1}{2}$ $\frac{19}{2}$

   (b) $6\frac{2}{3}$ $\frac{20}{3}$

   (c) $10\frac{5}{6}$ $\frac{65}{6}$

   (d) $8\frac{4}{9}$ $\frac{76}{9}$

2. Express each value as a whole number or mixed number in simplest form.

   (a) $\frac{25}{2}$ $12\frac{1}{2}$

   (b) $\frac{51}{7}$ $7\frac{2}{7}$

   (c) $\frac{46}{4}$ $11\frac{1}{2}$

   (d) $\frac{40}{3}$ $13\frac{1}{3}$

   (e) $46 \div 8$ $5\frac{3}{4}$

   (f) $85 \div 9$ $9\frac{4}{9}$

3. 45 pounds of rice is divided equally into 6 bags. How many pounds of rice are in each bag?
   $45 \div 6 = \frac{45}{6} = 7\frac{1}{2}$

   $7\frac{1}{2}$ pounds

4. Continue the patterns. Express numbers 1 or greater as whole or mixed numbers. Use simplest form.

   (a) Count on by three fourths starting with $\frac{1}{4}$.

   $\frac{1}{4}, 1, 1\frac{3}{4}, \boxed{2\frac{1}{2}}, \boxed{3\frac{1}{4}}, \boxed{4}, \boxed{4\frac{3}{4}}, \boxed{5\frac{1}{2}}, \boxed{6\frac{1}{4}}$

   (b) Count on by three eighths starting with $\frac{1}{8}$.

   $\frac{1}{8}, \boxed{\frac{1}{2}}, \boxed{\frac{7}{8}}, \boxed{1\frac{1}{4}}, \boxed{1\frac{5}{8}}, \boxed{2}, \boxed{2\frac{3}{8}}, \boxed{2\frac{3}{4}}$

   (c) Count on by two ninths starting with $\frac{2}{9}$.

   $\frac{2}{9}, \boxed{\frac{4}{9}}, \boxed{\frac{2}{3}}, \boxed{\frac{8}{9}}, \boxed{1\frac{1}{9}}, \boxed{1\frac{1}{3}}, \boxed{1\frac{5}{9}}, \boxed{1\frac{7}{9}}$

5. Write the numbers in order from least to greatest.

   (a) $\frac{16}{3}, 4\frac{7}{8}, \frac{26}{6}$

   $\frac{26}{6}, 4\frac{7}{8}, \frac{16}{3}$

   (b) $\frac{42}{10}, \frac{60}{9}, \frac{52}{7}, \frac{28}{5}$

   $\frac{42}{10}, \frac{28}{5}, \frac{60}{9}, \frac{52}{7}$

6. (a) Which fraction is closest to 5?

   $\frac{14}{3}$ $\boxed{\frac{19}{4}}$ $\frac{11}{5}$ $\frac{43}{8}$

   (b) Which fraction is closest to 10?

   $\boxed{\frac{59}{6}}$ $\frac{50}{4}$ $\frac{68}{7}$ $\frac{78}{8}$

7. Write the whole numbers that are between $\frac{19}{3}$ and $\frac{47}{4}$.
   7, 8, 9, 10, 11

Challenge

8. One half is one third of what mixed number?
   $1\frac{1}{2}$

   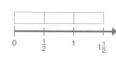

9. Two lumps of clay weigh the same. If you put 1 lump of clay on one side of a balance, and $\frac{3}{5}$ of the second lump of clay along with a $\frac{1}{2}$ kg weight on the other side of the balance, the two sides are balanced. How much does a lump of clay weigh?
   $\frac{2}{5}$ of each lump must be $\frac{1}{2}$ kg. If 2 fifths is half a kilogram, then 1 fifth is a fourth of a kilogram. 5 fifths is $\frac{5}{4}$ kg or $1\frac{1}{4}$ kg. Students may draw a picture.

Chapter 7 Adding and Subtracting Fractions

Overview

Suggested number of class periods: 7–8

	Lesson	Page	Resources	Objectives
	Chapter Opener	p. 215	TB: p. 161	Investigate adding and subtracting fractions.
1	Adding and Subtracting Fractions — Part 1	p. 216	TB: p. 162 WB: p. 151	Add and subtract fractions with like denominators.
2	Adding and Subtracting Fractions — Part 2	p. 218	TB: p. 165 WB: p. 154	Add and subtract fractions with related denominators.
3	Adding a Mixed Number and a Fraction	p. 220	TB: p. 168 WB: p. 157	Add a fraction and a mixed number with related denominators.
4	Adding Mixed Numbers	p. 223	TB: p. 171 WB: p. 160	Add mixed numbers with like or related denominators.
5	Subtracting a Fraction from a Mixed Number	p. 225	TB: p. 173 WB: p. 162	Subtract a fraction from a mixed number with like or related denominators.
6	Subtracting Mixed Numbers	p. 227	TB: p. 176 WB: p. 165	Subtract mixed numbers with like or related denominators.
7	Practice	p. 230	TB: p. 179 WB: p. 167	Practice adding and subtracting fractions.
	Workbook Solutions	p. 232		

Chapter 7 Adding and Subtracting Fractions

In Dimensions Math 3A, students added and subtracted fractions with like denominators when the total was less than or equal to 1. This concept is reviewed briefly before students move on to adding and subtracting both proper fractions and improper fractions, or fractions equal to or greater than 1.

Addition and Subtraction with Like Denominators

Similar to how whole numbers count like objects (4 bears, 3 cats, 7 animals), fractions can count like units. These units are equal sized parts of a whole number, such as sevenths, or fifths. Fractions of the same units have the same denominator and are sometimes called fractions with like denominators, or "like fractions."

To add or subtract fractions, the unit, or denominator, must be the same. We can easily add $\frac{5}{7}$ and $\frac{2}{7}$, but not $\frac{5}{7}$ and $\frac{5}{9}$, which requires taking additional steps. An analogy would be that while we can add 5 feet and 5 feet, we cannot add 5 feet and 5 meters without first converting to the same units. Similarly, to add fractions with different denominators, we need to make the units the same. We do so by finding equivalent fractions.

Addition and Subtraction with Unlike Denominators

"Unlike fractions" do not have the same denominator. In this chapter, students will learn to add fractions in which one denominator is a simple multiple of the other. For example, $\frac{3}{4}$ and $\frac{1}{2}$ are "related fractions" as 2 is a factor of 4 and 4 is a multiple of 2. In order to add related fractions, we must find an equivalent fraction for one of the fractions. In this chapter, when students are asked to add or subtract more than two fractions, one of the fractions will be related to at least one other fraction.

In Dimensions Math 5, students will learn to add "unrelated fractions," or fractions in which one denominator is not a multiple of the other.

For example:

$\frac{3}{5} + \frac{2}{3}$

In order to add or subtract related fractions, students will need to consider the fraction with the least denominator, and determine what that denominator can be multiplied by to arrive at an equivalent fraction with a like denominator of the other. For example, in $\frac{2}{3} + \frac{1}{6}$, 6 is a multiple of 3. 3 multiplied by 2 is 6, so students use this multiple to find an equivalent fraction for $\frac{2}{3}$ ($\frac{4}{6}$).

They can now add the fractions: $\frac{4}{6} + \frac{1}{6} = \frac{5}{6}$.

Similarly, $\frac{9}{10} - \frac{2}{5} = \frac{9}{10} - \frac{4}{10} = \frac{5}{10} = \frac{1}{2}$.

Note that in the second example, the final answer is simplified. Students should always simplify their answers because fractions with smaller denominators are easier to visualize, compare, and calculate.

The terms "like," "related," and "unlike" are not used with the students in the textbook. They are included in the Teacher's Guide for teacher reference.

Adding Mixed Numbers

To add a mixed number and a fraction, the mixed number can be thought of as the sum of a whole number and a fraction.

For example:

$1\frac{5}{6} + \frac{5}{6}$

We can add the fractional parts together:

$$1 + \frac{5}{6} + \frac{5}{6} = 1\frac{10}{6} = 2\frac{4}{6}$$

$$\frac{6}{6} \quad \frac{4}{6}$$

Since mixed numbers are not normally expressed with an improper fraction, this must be simplified.

$1 + \frac{6}{6} + \frac{4}{6} = 1 + 1 + \frac{4}{6}$
$\qquad = 2\frac{4}{6} = 2\frac{2}{3}$

Chapter 7 Adding and Subtracting Fractions

Notes

Students may also simplify the fraction first:

$$1 + \frac{5}{6} + \frac{5}{6} = 1 + \frac{10}{6}$$
$$= 1 + \frac{5}{3} = 2\frac{2}{3}$$

Although not taught in the textbook, students can use number bonds to make the next whole number in the same way they have used number bonds to make convenient numbers to add or subtract.

$$1\frac{5}{6} + \frac{5}{6} = 2\frac{4}{6} = 2\frac{2}{3}$$
$$\diagup \quad \diagdown$$
$$\frac{1}{6} \quad \frac{4}{6}$$

When adding mixed numbers, for example $3\frac{1}{2} + 1\frac{5}{8}$, students will be shown two methods.

Method 1

Students start by adding the first mixed number and the whole number of the second mixed number. Next, they add the remaining fractional part of the second number using concepts for adding related fractions and adding fractions to mixed numbers.

$$3\frac{1}{2} + 1\frac{5}{8}$$

$$3\frac{1}{2} \xrightarrow{+1} 4\frac{1}{2}$$

$$4\frac{1}{2} = 4\frac{4}{8}$$

$$4\frac{4}{8} \xrightarrow{+\frac{5}{8}} 4\frac{9}{8} = 5\frac{1}{8}$$

Method 2

Add the whole number parts and the fraction parts separately:

$$3 + \frac{1}{2} + 1 + \frac{5}{8} = 3 + 1 + \frac{1}{2} + \frac{5}{8}$$
$$= 4 + \frac{4}{8} + \frac{5}{8}$$
$$= 4\frac{9}{8} = 5\frac{1}{8}$$

Subtracting Mixed Numbers

In order to subtract a fraction from a mixed number when the fraction part of the mixed number is less than the fraction being subtracted, students will regroup 1 whole to a fraction in order to have enough parts from which to subtract.

For example:

$$2\frac{1}{6} - \frac{2}{3}$$

As $\frac{2}{3}$ is greater than $\frac{1}{6}$, we can think of 2 as 1 and $\frac{6}{6}$ and add the sixths together:

$$2\frac{1}{6} - \frac{2}{3} = 1\frac{7}{6} - \frac{4}{6}$$
$$\diagup \; |$$
$$1 \quad \frac{6}{6}$$

$\frac{2}{3}$ is converted to $\frac{4}{6}$ and now it is possible to subtract:

$$1\frac{7}{6} - \frac{4}{6} = 1\frac{3}{6} = 1\frac{1}{2}$$

A similar process is followed when subtracting one mixed number from another:

$$3\frac{3}{8} - 1\frac{3}{4}$$

Students can subtract the whole number first, and then the fraction, using concepts they have already learned in previous lessons.

$$3\frac{3}{8} \xrightarrow{-1} 2\frac{3}{8}$$

Convert the remaining part of the fraction to be subtracted, $\frac{3}{4}$, to eighths:

$$\frac{3}{4} = \frac{6}{8}$$

Chapter 7 Adding and Subtracting Fractions

Subtract $\frac{6}{8}$ from $2\frac{3}{8}$. Since $\frac{6}{8}$ is greater than $\frac{3}{8}$, we can think of $2\frac{3}{8}$ as $1 + \frac{8}{8} + \frac{3}{8}$ or $1\frac{11}{8}$.

$$2\frac{3}{8} - \frac{6}{8} = 1\frac{11}{8} - \frac{6}{8}$$
$$1 \quad \frac{8}{8} \qquad = 1\frac{5}{8}$$

Students may also find that it is easier to convert mixed numbers to improper fractions, subtract, and then simplify:

$3\frac{3}{8} - 1\frac{3}{4}$

$3\frac{3}{8} = \frac{27}{8}$

$1\frac{3}{4} = 1\frac{6}{8}$

$\qquad = \frac{14}{8}$

$\frac{27}{8} - \frac{14}{8} = \frac{13}{8}$

$\qquad\qquad = 1\frac{5}{8}$

Mixed number calculations should not be written in a vertical format similar to calculations with whole numbers. The purpose of a vertical algorithm for whole number calculations is to emphasize the single digit calculations within a base-ten system. Because fractions do not have a fixed base, a vertical algorithm is not used.

Materials

- Dry erase markers
- Fraction circles or bars
- Fraction manipulatives
- Pattern blocks
- Whiteboards

Blackline Masters

- Closest to 5 Fraction Cards
- Tower of Hanoi

Notes & Materials

Activities

Activities included in this chapter provide practice with fractions and mixed numbers. They can be used after students complete the **Do** questions or anytime additional practice is needed.

Chapter Opener

Objective

- Investigate adding and subtracting fractions.

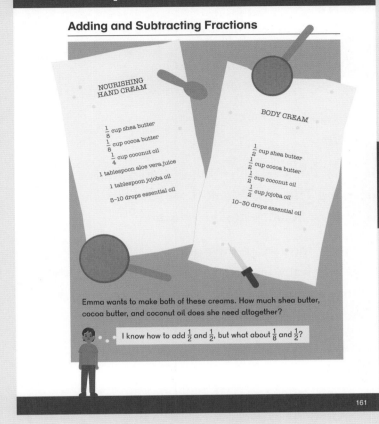

Discuss the examples in the **Chapter Opener** and Alex's question.

Have students recall what they know about adding fractions with the same denominator. Discuss cooking or other scenarios where students might need to add or subtract fractions either with the same or different denominators. Ask students to think about how to express addition and subtraction answers that are greater than one.

This lesson may continue straight to Lesson 1.

Lesson 1 Adding and Subtracting Fractions — Part 1

Objective
- Add and subtract fractions with like denominators.

Lesson Materials
- Fraction manipulatives

Think

Provide students with fraction manipulatives and pose the **Think** problem. Have students show or write their solutions to the questions.

Students may share their answers either as improper fractions or mixed numbers. They may choose to draw different representations, including a bar model or number line.

Discuss student solutions.

Learn

(a) Emma knows that both fractions name the same unit, fifths. She can simply add the numerators. Students can see this easily using manipulatives. For example, using fraction circles:

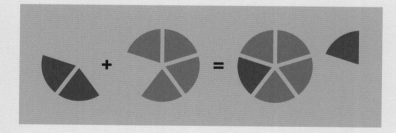

(b) Dion also sees fifths. He can simply subtract the numerators.

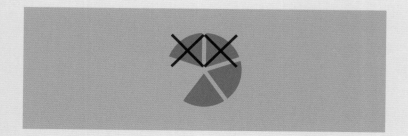

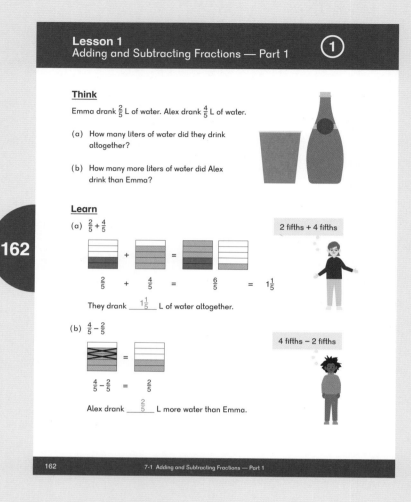

Have students compare their methods from **Think** with the ones in the textbook.

A common error is for students to add or subtract across the numerators and denominators. For example, students might mistakenly do the following:

$$\frac{2}{5} + \frac{4}{5} = \frac{6}{10}$$

$$\frac{4}{5} - \frac{2}{5} = \frac{2}{0}$$

Manipulatives help students see that the denominator remains fifths when adding and subtracting fifths.

Do

1 — 2 Students should give their answers in simplest form.

5 — 6 Have students share how they solved some of these problems.

Activity

▲ Decomposing Fractions

Have students find different ways to decompose fractions.

For example, $\frac{7}{8}$ can be decomposed to:

- $\frac{6}{8} + \frac{1}{8}$
- $\frac{5}{8} + \frac{2}{8}$
- $\frac{4}{8} + \frac{3}{8}$
- $\frac{5}{8} + \frac{1}{8} + \frac{1}{8}$
- $\frac{2}{8} + \frac{2}{8} + \frac{3}{8}$

Exercise 1 • page 151

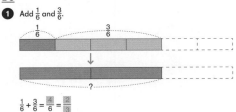

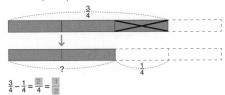

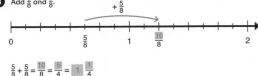

Lesson 2 Adding and Subtracting Fractions — Part 2

Objective

- Add and subtract fractions with related denominators.

Lesson Materials

- Fraction manipulatives

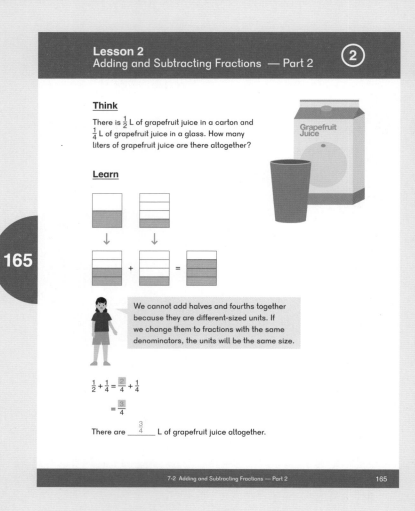

Think

Provide students with fraction manipulatives and pose the **Think** problem. Have students show or write their solutions to the questions.

Students should see that halves and fourths are different sized units. In order to add, we need to make equal fractional units:

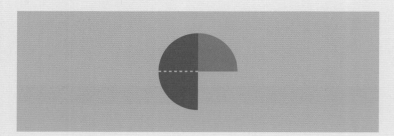

Learn

Discuss Mei's comment.

To add the fractions, we need to use the same units. We can express $\frac{1}{2}$ as an equivalent fraction with a denominator of fourths: $\frac{2}{4}$.

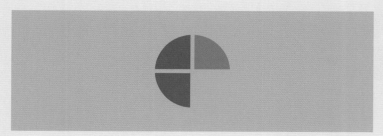

Now that both fractions are expressed in the same unit, fourths, they can be added.

Have students compare their solution from **Think** with the one in the textbook.

Do

1. The two top bars show $\frac{1}{5}$ is the same size as $\frac{2}{10}$. The bottom bar shows that we can now add the equal sized units.

   Finally, the answer, $\frac{5}{10}$, is simplified to $\frac{1}{2}$. Students might see that the denominator of the final, simplified answer is not the same as the denominator of either of the original fractions being added.

2. The top bars show that $\frac{1}{2}$ is the same size as $\frac{3}{6}$. The bottom bar shows the result when $\frac{3}{6}$ is taken away from $\frac{5}{6}$.

   Students should understand that the 3 boxes that are crossed out are equivalent to the $\frac{1}{2}$ that is subtracted.

   Dion reminds students to make equal units.

3. — 4. Students first need to find an equivalent fraction. The improper fractions are added or subtracted and then simplified. In ③, the final answer is converted to a mixed number.

5. Have students share how they solved some of these problems, specifically (h) and (i), which are new concepts.

   Extend by asking students how they would solve a problem with three different denominators such as: $\frac{5}{12} + \frac{1}{3} + \frac{1}{2}$.

6. Have students draw a bar model, as needed, to compare the numbers and find the difference.

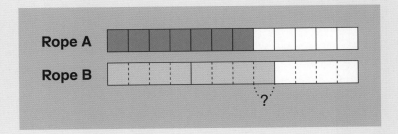

Exercise 2 • page 154

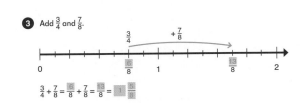

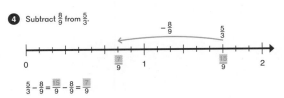

Lesson 3 Adding a Mixed Number and a Fraction

Objective

- Add a fraction and a mixed number with related denominators.

Lesson Materials

- Fraction circles or bars

Think

Provide students with fraction circles or bars and pose the **Think** problem.

Ask students how this problem is different from the ones they have done in previous lessons.

Have students show or write their solutions to the questions.

Learn

Students should see that they do not need to change $1\frac{3}{4}$ to an improper fraction to solve the problem, but can simply add the proper fractions together.

Since $1\frac{3}{4} + \frac{3}{4} = 1 + \frac{3}{4} + \frac{3}{4}$, and since addition can be done in any order, students can add the fractions together, then express $\frac{6}{4}$ as $1\frac{1}{2}$, and find the total $2\frac{1}{2}$.

Students may have thought of a number bond and split $\frac{3}{4}$ into $\frac{1}{4}$ and $\frac{2}{4}$:

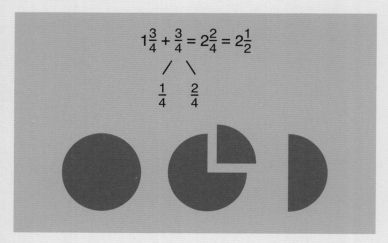

Have students compare their methods from **Think** with the one in the textbook.

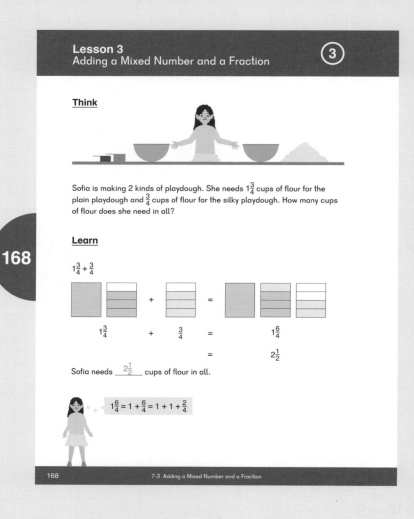

Do

1 – 2 Students will first need to make equal fractional units.

2 There are other solutions that students might see.

Students could first simplify $1\frac{15}{12}$ to $1\frac{5}{4}$, then convert to $2\frac{1}{4}$.

Students could also add to make the next whole by splitting $\frac{7}{12}$ into $\frac{4}{12}$ and $\frac{3}{12}$:

$$1\frac{8}{12} + \frac{7}{12} = 2\frac{3}{12} = 2\frac{1}{4}$$
$$\diagup \quad \diagdown$$
$$\frac{4}{12} \quad \frac{3}{12}$$

3 Students can add to make the next whole by splitting $\frac{5}{6}$ into $\frac{1}{6}$ and $\frac{4}{6}$:

$$1\frac{5}{6} + \frac{5}{6} = 2\frac{4}{6} = 2\frac{2}{3}$$
$$\diagup \quad \diagdown$$
$$\frac{1}{6} \quad \frac{4}{6}$$

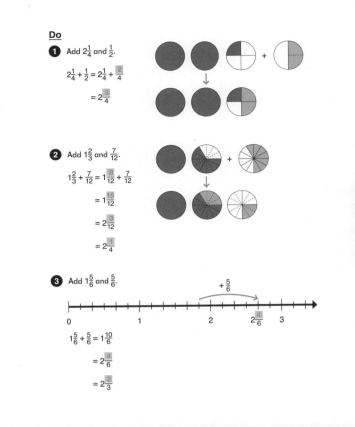

4. Students can add to make the next whole by splitting $\frac{6}{8}$ into $\frac{3}{8}$ and $\frac{3}{8}$:

$$1\frac{5}{8} + \frac{6}{8} = 1 + \frac{8}{8} + \frac{3}{8}$$
$$\diagup \quad \diagdown$$
$$\frac{3}{8} \quad \frac{3}{8}$$

5. Students can add to make the next whole by splitting $\frac{5}{6}$ into $\frac{2}{6}$ and $\frac{3}{6}$:

$$2\frac{4}{6} + \frac{5}{6} = 2 + \frac{6}{6} + \frac{3}{6}$$
$$\diagup \quad \diagdown$$
$$\frac{2}{6} \quad \frac{3}{6}$$

6. Have students share how they solved some of these problems. Note that problems (g)–(i) might generate more discussion.

Activity

▲ Pattern Blocks Fractions

Materials: Pattern blocks

Provide students with pattern blocks. Name one piece as 1 whole and then have them practice by adding the values of the blocks.

For example:

- If the trapezoid is 1 whole, find the value of 1 hexagon and 1 green triangle. $2\frac{1}{3}$
- If the trapezoid is 1 whole, find the value of 1 hexagon and 4 green triangles. $3\frac{1}{3}$
- If the hexagon is 1 whole, find the value of 1 trapezoid and 4 green triangles. $1\frac{1}{6}$

Have students make different equations with the pattern blocks.

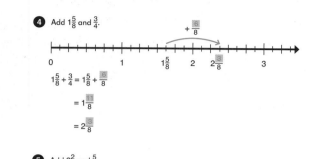

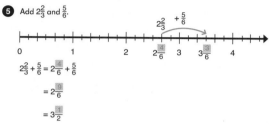

Exercise 3 · page 157

222   Teacher's Guide 4A Chapter 7   © 2019 Singapore Math Inc.

Lesson 4 Adding Mixed Numbers

Objective
- Add mixed numbers with like or related denominators.

Lesson Materials
- Fraction manipulatives

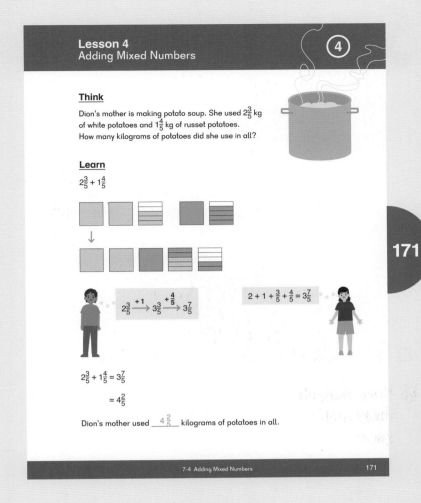

Think

Provide students with fraction manipulatives and pose the **Think** problem. Ask students how this problem is different from ones they have done in previous lessons. Students should see that both numbers are mixed numbers.

Have students show or write their solutions to the questions.

Learn

Discuss the two methods shown in **Learn**. Alex thinks of $1\frac{4}{5}$ as $1 + \frac{4}{5}$ and adds the whole number to $2\frac{3}{5}$ first.

Mei adds the whole numbers together first and then adds the fractions together.

Mei knows $2 + \frac{3}{5} + 1 + \frac{4}{5} = 2 + 1 + \frac{3}{5} + \frac{4}{5}$.

Students may also have solved the problem by thinking about number bonds:

$$2\frac{3}{5} + 1\frac{4}{5} = 3 + 1\frac{2}{5} \qquad 2\frac{3}{5} + 1\frac{4}{5} = 2\frac{2}{5} + 2$$
$$\diagup \; \diagdown \qquad\qquad \diagup \; \diagdown$$
$$\frac{2}{5} \quad 1\frac{2}{5} \qquad\qquad 2\frac{2}{5} \quad \frac{1}{5}$$

Students may also convert the mixed numbers to improper fractions and then add, although this is less likely to happen when fraction manipulatives are used:

$$2\frac{3}{5} + \frac{9}{5} = \frac{13}{5} + \frac{9}{5} = \frac{22}{5} = 4\frac{2}{5}$$

Do

① To add, students will first need to make equal fractional units for the fraction part of the mixed numbers.

They add the whole number first, and then the fractions.

Students could also add to make the next whole:

$$3\frac{5}{6} + \frac{4}{6} = 4\frac{3}{6} = 4\frac{1}{2}$$
$$\frac{1}{6} \quad \frac{3}{6}$$

② Once students find common denominators, they can add the whole number first, and then the fractions.

Students could also add to make the next whole:

$$8\frac{6}{9} + \frac{5}{9} = 9\frac{2}{9}$$
$$\frac{3}{9} \quad \frac{2}{9}$$

③ Have students share how they solved some of these problems.

Exercise 4 • page 160

Lesson 5 Subtracting a Fraction from a Mixed Number

Objective

- Subtract a fraction from a mixed number with like or related denominators.

Lesson Materials

- Fraction manipulatives

Think

Provide students with fraction manipulatives and pose the **Think** problem.

Discuss student solutions to the questions.

Learn

Students begin with fraction manipulatives that show 2 wholes and $\frac{1}{5}$.

They can easily see that in order to subtract $\frac{4}{5}$, they need to change 1 whole into fractional units. $\frac{5}{5}$ added to the $\frac{1}{5}$ makes $\frac{6}{5}$.

Students can then subtract the $\frac{4}{5}$.

$2\frac{1}{5} - \frac{4}{5} = 1\frac{6}{5} - \frac{4}{5} = 1\frac{2}{5}$

Discuss the example in the textbook as well as Emma's comment. She thinks of $2\frac{1}{5}$ as $1 + 1 + \frac{1}{5}$, and converts 1 into fractional units, $\frac{5}{5}$.

Students could also have subtracted from 1 whole:

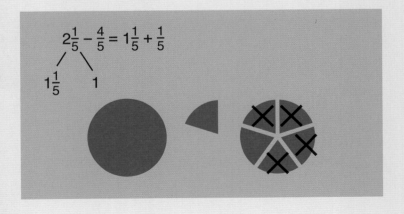

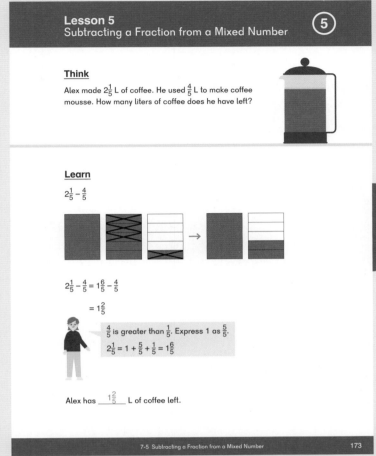

A common error is for students to switch the fractional units before subtracting:

$\frac{4}{5} - \frac{1}{5}$

$2 + \frac{1}{5} - \frac{4}{5} = 2\frac{4}{5} - \frac{1}{5} = 2\frac{3}{5}$

Remind students that $2\frac{1}{5}$ is the whole and we are subtracting a part:

$2\frac{1}{5} - \frac{4}{5}$ is the same as $2\frac{1}{5}$.

Do

① Note the progression of the problems from (a) to (c). (a) subtracts from 1 whole, (b) subtracts from 2 wholes, and (c) subtracts from a mixed number. Have students discuss the similarities and differences among the problems and the steps necessary to solve each one.

②—④ To subtract, students will first need to express the fractional units as fractions with like denominators and then change one whole into fractional units.

⑤ Have students share how they solved some of these problems.

Problems (b)–(d) involve regrouping from the whole number.

In problems (e) and (f), there is no regrouping. Students only need to make equal fractional units.

Problems (g)–(i) involve making equal fractional units and regrouping.

Exercise 5 • page 162

Do

① (a) Subtract $\frac{5}{7}$ from 1.

$1 - \frac{5}{7} = \frac{7}{7} - \frac{5}{7}$

$= \frac{2}{7}$

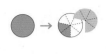

(b) Subtract $\frac{5}{7}$ from 2.

$2 - \frac{5}{7} = 1\frac{7}{7} - \frac{5}{7}$

$= 1\frac{2}{7}$

(c) Subtract $\frac{5}{7}$ from $2\frac{3}{7}$.

$2\frac{3}{7} - \frac{5}{7} = 1\frac{10}{7} - \frac{5}{7}$

$= 1\frac{5}{7}$

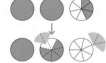

② Subtract $\frac{2}{3}$ from $3\frac{1}{6}$.

$3\frac{1}{6} - \frac{2}{3} = 3\frac{1}{6} - \frac{4}{6}$

$= 2\frac{7}{6} - \frac{4}{6}$

$= 2\frac{3}{6}$

$= 2\frac{1}{2}$

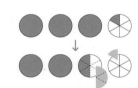

③ Subtract $\frac{5}{8}$ from $1\frac{1}{2}$.

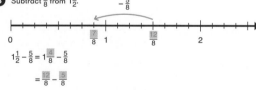

$1\frac{1}{2} - \frac{5}{8} = 1\frac{4}{8} - \frac{5}{8}$

$= \frac{12}{8} - \frac{5}{8}$

$= \frac{7}{8}$

④ $6\frac{5}{12} - \frac{3}{4} = 6\frac{5}{12} - \frac{9}{12}$

$= 5\frac{17}{12} - \frac{9}{12}$

$= 5\frac{8}{12}$

$= 5\frac{2}{3}$

⑤ Subtract. Express each answer in its simplest form.

(a) $7\frac{5}{11} - \frac{3}{11}$ $7\frac{2}{11}$ (b) $3\frac{1}{9} - \frac{5}{9}$ $2\frac{5}{9}$ (c) $5\frac{3}{10} - \frac{7}{10}$ $4\frac{3}{5}$

(d) $6\frac{5}{8} - \frac{7}{8}$ $5\frac{3}{4}$ (e) $7\frac{9}{14} - \frac{2}{7}$ $7\frac{5}{14}$ (f) $4\frac{3}{4} - \frac{7}{12}$ $4\frac{1}{6}$

(g) $9\frac{3}{10} - \frac{1}{2}$ $8\frac{4}{5}$ (h) $6\frac{1}{2} - \frac{7}{8}$ $5\frac{5}{8}$ (i) $8\frac{1}{3} - \frac{4}{9}$ $7\frac{8}{9}$

Exercise 5 • page 162

Lesson 6 Subtracting Mixed Numbers

Objective
- Subtract mixed numbers with like or related denominators.

Lesson Materials
- Fraction manipulatives

Think

Provide students with fraction manipulatives and pose the **Think** problem.

Discuss student solutions to the questions.

Learn

Discuss Sophia's thought. She subtracts the whole number from the mixed number, then subtracts the fractional part of the second mixed number from that difference.

In addition to the example shown in the textbook, students could subtract the whole numbers and the fractions separately, then add the two differences. That strategy does not always work when there is subtraction with regrouping.

Sofia subtracts the whole number first:

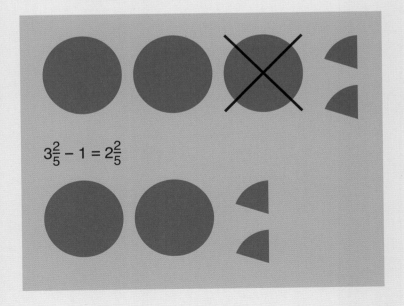

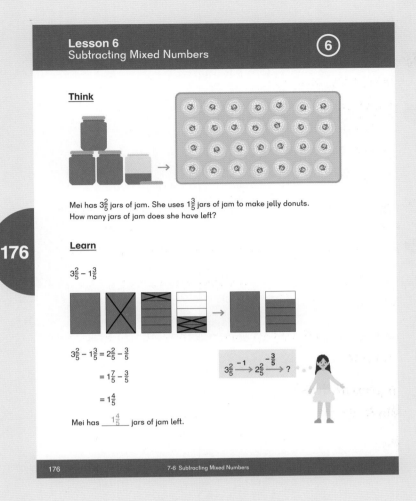

1 whole can then be regrouped into fractional parts, $\frac{5}{5}$. There are now $\frac{7}{5}$ and Sofia subtracts $\frac{3}{5}$ from $\frac{7}{5}$:

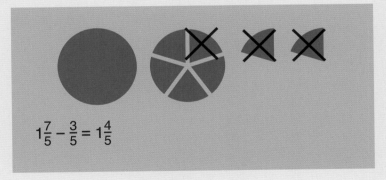

Do

❶—❺ Discuss the steps given in each problem with students.

❷ Subtract the 2 wholes first.

Regroup 1 whole as $\frac{5}{5}$.

Subtract: $\frac{5}{5} - \frac{1}{5}$.

$1\frac{5}{5} - \frac{1}{5} = 1\frac{4}{5}$

❸ Subtract the 1 whole first.

Change the fractional parts to make equal fractional units: $1\frac{3}{8} - \frac{6}{8}$.

Now students should see that they need to regroup 1 whole as $\frac{8}{8}$.

They add that to the $\frac{3}{8}$ to make $\frac{11}{8}$.

Subtract: $\frac{11}{8} - \frac{6}{8}$.

④ Subtract the 1 whole first.

Regroup 1 whole as $\frac{3}{3}$, then add that to the $\frac{1}{3}$ to make $5\frac{4}{3}$.

Subtract: $5\frac{4}{3} - \frac{2}{3}$.

⑤ Subtract the 4 wholes first.

Change the fractional parts to make equal fractional units: $\frac{1}{2} = \frac{3}{6}$.

Regroup 1 whole as $\frac{6}{6}$, then add that to the $\frac{1}{6}$ to make $\frac{7}{6}$.

Subtract: $4\frac{7}{6} - \frac{3}{6} = 4\frac{4}{6}$.

Simplify $4\frac{4}{6}$ to $4\frac{2}{3}$.

⑥ Have students share how they solved some of these problems.

Example solutions for (g)–(o):

(g) $6\frac{2}{5} - 4\frac{4}{5} = 2\frac{2}{5} - \frac{4}{5}$
$= 1\frac{7}{5} - \frac{4}{5} = 1\frac{3}{5}$

(h) $9\frac{3}{8} - 3\frac{7}{8} = 6\frac{3}{8} - \frac{7}{8}$
$= 5\frac{11}{8} - \frac{7}{8} = 5\frac{4}{8} = 5\frac{1}{2}$

(i) $8\frac{5}{9} - 5\frac{7}{9} = 3\frac{5}{9} - \frac{7}{9}$
$= 2\frac{14}{9} - \frac{7}{9} = 2\frac{7}{9}$

In problems (j)–(l) there is no regrouping needed.

(m) $7\frac{1}{3} - 2\frac{9}{12} = 5\frac{1}{3} - \frac{9}{12}$
$= 5\frac{4}{12} - \frac{9}{12}$
$= 4\frac{16}{12} - \frac{9}{12} = 4\frac{7}{12}$

(n) $6\frac{1}{4} - 3\frac{7}{8} = 3\frac{1}{4} - \frac{7}{8}$
$= 3\frac{2}{8} - \frac{7}{8}$
$= 2\frac{10}{8} - \frac{7}{8} = 2\frac{3}{8}$

(o) $8\frac{4}{9} - 7\frac{2}{3} = 1\frac{4}{9} - \frac{2}{3}$
$= 1\frac{4}{9} - \frac{6}{9}$
$= \frac{13}{9} - \frac{6}{9} = \frac{7}{9}$

Exercise 6 • page 165

Lesson 7 Practice

Objective

- Practice adding and subtracting fractions.

After students complete the **Practice** in the textbook, have them continue to practice adding and subtracting mixed numbers with activities from the chapter.

Activity

▲ Closest to 5

Materials: Closest to 5 Fraction Cards (BLM)

Shuffle the cards and deal five to each player.

Players select two of their cards and add the fractions together. The player whose sum is closest to 5 earns a point.

For example:

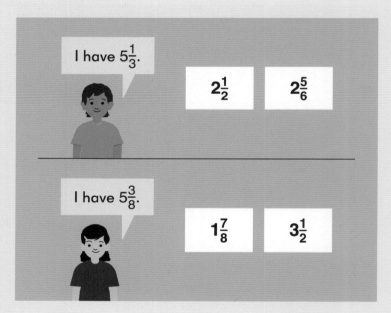

In this example, Alex is closer to 5 than Mei. He earns 1 point. Players draw two new cards for the next round. The first player to earn 5 points is the winner.

Lesson 7 Practice P 7

1. Add or subtract. Express each answer in its simplest form.

   (a) $\frac{4}{9} + \frac{8}{9}$ $1\frac{1}{3}$
   (b) $\frac{11}{12} - \frac{7}{12}$ $\frac{1}{3}$
   (c) $\frac{3}{7} + \frac{5}{7}$ $1\frac{1}{7}$
   (d) $\frac{2}{3} - \frac{2}{9}$ $\frac{4}{9}$
   (e) $\frac{1}{6} + \frac{3}{12}$ $\frac{5}{12}$
   (f) $\frac{5}{6} - \frac{2}{3}$ $\frac{1}{6}$
   (g) $\frac{13}{16} - \frac{3}{8}$ $\frac{7}{16}$
   (h) $\frac{17}{20} + \frac{3}{5}$ $1\frac{9}{20}$
   (i) $\frac{9}{10} + \frac{1}{2} + \frac{3}{5}$ 2

2. Add or subtract. Express each answer in its simplest form.

   (a) $12 - 7\frac{5}{9}$ $4\frac{4}{9}$
   (b) $5\frac{1}{8} + 3\frac{3}{8}$ $8\frac{1}{2}$
   (c) $4\frac{5}{12} + 5\frac{7}{12}$ 10
   (d) $4\frac{3}{5} - 1\frac{1}{10}$ $3\frac{1}{2}$
   (e) $9\frac{5}{8} - 3\frac{1}{4}$ $6\frac{3}{8}$
   (f) $8\frac{2}{3} - 5\frac{2}{9}$ $3\frac{4}{9}$
   (g) $7\frac{1}{6} - 4\frac{1}{2}$ $2\frac{2}{3}$
   (h) $9\frac{3}{8} - 3\frac{3}{4}$ $5\frac{5}{8}$
   (i) $10\frac{2}{3} - 4\frac{5}{6}$ $5\frac{5}{6}$

3. Amanda ran $\frac{7}{10}$ km on Monday, the same distance on Tuesday, and $\frac{9}{10}$ km on Wednesday. How far did she run altogether?
   $\frac{7}{10} + \frac{7}{10} + \frac{9}{10} = 2\frac{3}{10}$; $2\frac{3}{10}$ km

4. Max and Lincoln are painting a fence. Max painted $\frac{1}{3}$ of it and Lincoln painted $\frac{4}{9}$ of it.

   (a) How much of the fence did they paint altogether?
   $\frac{1}{3} + \frac{4}{9} = \frac{7}{9}$; $\frac{7}{9}$ of the fence
   (b) How much more of the fence did Lincoln paint than Max?
   $\frac{4}{9} - \frac{1}{3} = \frac{1}{9}$; $\frac{1}{9}$ more

Brain Works

★ **Tower of Hanoi**

Materials: 5 circles of different sizes or Tower of Hanoi (BLM)

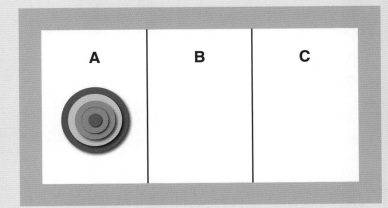

Stack the circles in order according to size, with the smallest circle on top in Section A. Move the circles from Section A to Section C using the following rules:

- Only one circle can be moved at a time.
- A circle can only be placed on an empty space or on a larger circle.

Exercise 7 • page 167

Exercise 1 • pages 151–153

Chapter 7 Adding and Subtracting Fractions

Exercise 1

Basics

1 (a) Add $\frac{5}{8}$ and $\frac{7}{8}$.

$\frac{5}{8} + \frac{7}{8} = \boxed{\frac{12}{8}}$

$= 1\boxed{\frac{4}{8}}$

$= 1\boxed{\frac{1}{2}}$

(b) Subtract $\frac{5}{8}$ from $\frac{7}{8}$.

$\frac{7}{8} - \frac{5}{8} = \boxed{\frac{2}{8}}$

$= \boxed{\frac{1}{4}}$

2 Add $\frac{7}{12}$ and $\frac{7}{12}$.

$\frac{7}{12} + \frac{7}{12} = \boxed{\frac{14}{12}} = \boxed{\frac{7}{6}} = 1\boxed{\frac{1}{6}}$

3 Subtract $\frac{7}{9}$ from $\frac{13}{9}$. Express the answer in simplest form.

$\frac{13}{9} - \frac{7}{9} = \boxed{\frac{2}{3}}$

Practice

Express answers 1 or greater as whole or mixed numbers. Use simplest form.

4 Add or subtract.

(a) $\frac{1}{12} + \frac{5}{12}$ $\quad \frac{1}{2}$

(b) $\frac{4}{8} + \frac{5}{8}$ $\quad 1\frac{1}{8}$

(c) $\frac{9}{10} - \frac{5}{10}$ $\quad \frac{2}{5}$

(d) $\frac{7}{10} + \frac{9}{10}$ $\quad 1\frac{3}{5}$

(e) $\frac{11}{13} - \frac{4}{13}$ $\quad \frac{7}{13}$

(f) $\frac{11}{15} + \frac{13}{15}$ $\quad 1\frac{3}{5}$

(g) $\frac{15}{8} - \frac{11}{8}$ $\quad \frac{1}{2}$

(h) $\frac{23}{20} + \frac{17}{20}$ $\quad 2$

(i) $\frac{4}{25} + \frac{9}{25} + \frac{7}{25}$ $\quad \frac{4}{5}$

(j) $\frac{23}{30} - \frac{17}{30}$ $\quad \frac{1}{5}$

5 A bag of rice weighs $\frac{13}{16}$ lb. A bag of wheat berries weighs $\frac{9}{16}$ lb more than the bag of rice. A bag of rye berries weighs $\frac{11}{16}$ lb less than the bag of wheat berries. What is the total weight of all three bags of grain?

Wheat berries: $\frac{13}{16} + \frac{9}{16} = \frac{22}{16}$

Rye berries: $\frac{22}{16} - \frac{11}{16} = \frac{11}{16}$

Total: $\frac{13}{16} + \frac{22}{16} + \frac{11}{16} = 2\frac{7}{8}$

Challenge

6 Write the missing digits. The fractions should be less than 1.

(a) $\frac{5}{6} + \frac{\boxed{5}}{\boxed{6}} = 1\frac{\boxed{2}}{\boxed{3}}$

(b) $\frac{\boxed{13}}{\boxed{15}} - \frac{\boxed{8}}{\boxed{15}} = \frac{\boxed{1}}{\boxed{3}}$

Exercise 2 • pages 154–156

Exercise 2

Basics

1. (a) Add $\frac{3}{4}$ and $\frac{7}{12}$.

   $\frac{3}{4} + \frac{7}{12} = \boxed{\frac{9}{12}} + \frac{7}{12}$

   $= \boxed{\frac{16}{12}}$

   $= 1\boxed{\frac{4}{12}}$

   $= 1\boxed{\frac{1}{3}}$

   (b) Subtract $\frac{7}{12}$ from $\frac{3}{4}$.

   $\frac{3}{4} - \frac{7}{12} = \boxed{\frac{9}{12}} - \frac{7}{12}$

   $= \boxed{\frac{2}{12}}$

   $= \boxed{\frac{1}{6}}$

2. Add $\frac{2}{3}$ and $\frac{5}{6}$.

   $\frac{2}{3} + \frac{5}{6} = \boxed{\frac{4}{6}} + \frac{5}{6} = \boxed{\frac{9}{6}} = \boxed{\frac{3}{2}} = 1\boxed{\frac{1}{2}}$

3. Subtract $\frac{11}{15}$ from $\frac{4}{3}$. Express the answer in simplest form.

   $\frac{4}{3} - \frac{11}{15} = \boxed{\frac{20}{15}} - \frac{11}{15} = \boxed{\frac{9}{15}} = \boxed{\frac{3}{5}}$

Practice

Express answers 1 or greater as whole or mixed numbers. Use simplest form.

4. Add.

   (a) $\frac{1}{2} + \frac{3}{14}$ $\quad \frac{5}{7}$

   (b) $\frac{5}{12} + \frac{3}{4}$ $\quad 1\frac{1}{6}$

   (c) $\frac{9}{8} + \frac{11}{24}$ $\quad 1\frac{7}{12}$

   (d) $\frac{2}{3} + \frac{7}{9}$ $\quad 1\frac{4}{9}$

   (e) $\frac{4}{5} + \frac{3}{10} + \frac{7}{10}$ $\quad 1\frac{4}{5}$

   (f) $\frac{4}{9} + \frac{17}{36} + \frac{11}{18}$ $\quad 1\frac{19}{36}$

5. Subtract.

   (a) $\frac{11}{14} - \frac{4}{7}$ $\quad \frac{3}{14}$

   (b) $\frac{5}{6} - \frac{5}{12}$ $\quad \frac{5}{12}$

   (c) $\frac{7}{5} - \frac{4}{15}$ $\quad 1\frac{2}{15}$

   (d) $\frac{8}{9} - \frac{2}{3}$ $\quad \frac{2}{9}$

   (e) $\frac{24}{25} - \frac{2}{5} - \frac{7}{50}$ $\quad \frac{21}{50}$

   (f) $\frac{11}{12} - \frac{2}{3} - \frac{1}{4}$ $\quad 0$

6. $\frac{3}{10}$ of a pole is painted red, $\frac{1}{5}$ of the pole is painted yellow, and the rest of the pole is painted blue. What fraction of the pole is painted blue?

   $\frac{10}{10} - \frac{3}{10} - \frac{1}{5} = \frac{10}{10} - \frac{3}{10} - \frac{2}{10}$

   $= \frac{5}{10}$

   $\frac{1}{2}$ of the pole

Exercise 3 • pages 157–159

Exercise 3

Basics

1 Add $1\frac{2}{3}$ and $\frac{7}{9}$.

$1\frac{2}{3} + \frac{7}{9} = 1\boxed{\frac{6}{9}} + \frac{7}{9}$

$= 1\boxed{\frac{13}{9}}$

$= 2\boxed{\frac{4}{9}}$

2 Add $1\frac{1}{2}$ and $\frac{5}{6}$.

$1\frac{1}{2} + \frac{5}{6} = 1\boxed{\frac{3}{6}} + \frac{5}{6} = 1\boxed{\frac{8}{6}} = 1\boxed{\frac{4}{3}} = 2\boxed{\frac{1}{3}}$

3 Add $\frac{3}{4}$ and $7\frac{5}{12}$.

$\frac{3}{4} + 7\frac{5}{12} = \boxed{\frac{9}{12}} + 7\frac{5}{12} = 7\boxed{\frac{14}{12}} = 7\boxed{\frac{7}{6}} = 8\boxed{\frac{1}{6}}$

Practice

Express answers 1 or greater as whole or mixed numbers. Use simplest form.

4 Add.

(a) $4\frac{3}{4} + \frac{3}{8}$ $5\frac{1}{8}$

(b) $6\frac{1}{2} + \frac{5}{12}$ $6\frac{11}{12}$

(c) $\frac{4}{9} + 4\frac{2}{3}$ $5\frac{1}{9}$

(d) $8\frac{4}{5} + \frac{7}{20}$ $9\frac{3}{20}$

(e) $\frac{11}{14} + 2\frac{4}{7}$ $3\frac{5}{14}$

(f) $\frac{5}{6} + 5\frac{13}{18}$ $6\frac{5}{9}$

(g) $\frac{1}{6} + \frac{7}{30} + 3\frac{1}{3}$ $3\frac{11}{15}$

(h) $\frac{2}{3} + 6\frac{3}{4} + \frac{7}{12}$ 8

5 Aurora ran $3\frac{3}{4}$ miles on Saturday. On Sunday, she ran $\frac{5}{12}$ miles farther than on Saturday. How far did she run altogether?

Sunday: $3\frac{3}{4} + \frac{5}{12} = 4\frac{1}{6}$

Total: $3\frac{3}{4} + 4\frac{1}{6} = 7\frac{11}{12}$

$7\frac{11}{12}$ miles

Challenge

6 Complete each problem using the given digits. Fractions should all be in simplest form.

(a) 12, 13, 14, 15

$\boxed{\frac{13}{24}} + 14\boxed{\frac{7}{8}} = 15\boxed{\frac{5}{12}}$

(b) 1, 3, 4, 5, 8

$4\boxed{\frac{2}{3}} + \boxed{\frac{8}{15}} = 5\boxed{\frac{1}{5}}$

Exercise 4 • pages 160–161

Exercise 4

Basics

1. Add.

   (a) $6\frac{1}{4} + 7\frac{7}{12} = 13\frac{1}{4} + \frac{7}{12}$
   $= 13\boxed{\frac{3}{12}} + \frac{7}{12}$
   $= 13\boxed{\frac{10}{12}}$
   $= 13\boxed{\frac{5}{6}}$

   (b) $4\frac{11}{24} + 2\frac{7}{8} = 6\frac{11}{24} + \frac{7}{8}$
   $= 6\frac{11}{24} + \boxed{\frac{21}{24}}$
   $= 6\boxed{\frac{32}{24}}$
   $= 7\boxed{\frac{8}{24}}$
   $= 7\boxed{\frac{1}{3}}$

Practice

2. Add. Express each answer as a mixed number in simplest form.

   (a) $7\frac{2}{9} + 4\frac{1}{3}$ $11\frac{5}{9}$

   (b) $6\frac{1}{2} + 2\frac{5}{14}$ $8\frac{6}{7}$

   (c) $4\frac{1}{3} + 3\frac{1}{6}$ $7\frac{1}{2}$

   (d) $8\frac{3}{4} + 4\frac{3}{20}$ $12\frac{9}{10}$

   (e) $1\frac{9}{12} + 3\frac{2}{3}$ $5\frac{5}{12}$

   (f) $5\frac{1}{2} + 4\frac{5}{8}$ $10\frac{1}{8}$

   (g) $6\frac{5}{6} + 2\frac{1}{18}$ $8\frac{8}{9}$

   (h) $3\frac{3}{4} + 3\frac{7}{12}$ $7\frac{1}{3}$

   (i) $4\frac{1}{3} + 2\frac{5}{12} + 3\frac{1}{3}$ $10\frac{1}{12}$

   (j) $1\frac{1}{2} + 6\frac{3}{4} + 5\frac{7}{8}$ $14\frac{1}{8}$

3. After using $2\frac{5}{8}$ yards of fabric to make a dress, Gina still had $7\frac{3}{4}$ yards of fabric left. How many yards of fabric did she have at first?
   $2\frac{5}{8} + 7\frac{3}{4} = 10\frac{3}{8}$

   $10\frac{3}{8}$ yards

Exercise 5 • pages 162–164

Exercise 5

Basics

1. Subtract $\frac{7}{12}$ from $2\frac{1}{4}$.

   $2\frac{1}{4} - \frac{7}{12} = 2\boxed{\frac{3}{12}} - \frac{7}{12}$
   $= 1\boxed{\frac{15}{12}} - \frac{7}{12}$
   $= 1\boxed{\frac{8}{12}}$
   $= 1\boxed{\frac{2}{3}}$

   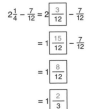

2. Subtract $\frac{5}{6}$ from $2\frac{1}{2}$.

   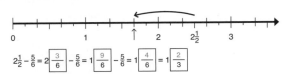

   $2\frac{1}{2} - \frac{5}{6} = 2\boxed{\frac{3}{6}} - \frac{5}{6} = 1\boxed{\frac{9}{6}} - \frac{5}{6} = 1\boxed{\frac{4}{6}} = 1\boxed{\frac{2}{3}}$

3. Subtract $\frac{5}{8}$ from $5\frac{5}{16}$.

   $5\frac{5}{16} - \frac{5}{8} = 5\frac{5}{16} - \boxed{\frac{10}{16}} = 4\boxed{\frac{21}{16}} - \boxed{\frac{10}{16}} = 4\boxed{\frac{11}{16}}$

Practice

Express all answers as mixed numbers in simplest form.

4. Subtract.

   (a) $4\frac{3}{4} - \frac{3}{8}$ $4\frac{3}{8}$

   (b) $4\frac{8}{9} - \frac{2}{3}$ $4\frac{2}{9}$

   (c) $6\frac{1}{2} - \frac{7}{12}$ $5\frac{11}{12}$

   (d) $8\frac{4}{5} - \frac{7}{20}$ $8\frac{9}{20}$

   (e) $9\frac{5}{14} - \frac{4}{7}$ $8\frac{11}{14}$

   (f) $3\frac{1}{6} - \frac{13}{18}$ $2\frac{4}{9}$

   (g) $8\frac{1}{6} - \frac{7}{30} - \frac{1}{3}$ $7\frac{3}{5}$

   (h) $9\frac{2}{3} - \frac{3}{4} - \frac{7}{12}$ $8\frac{1}{3}$

5. Andrei spent $3\frac{1}{2}$ hours working in his garden on Saturday. He spent $\frac{3}{4}$ of an hour less time working in his garden on Sunday than on Saturday. How much time did he spend working in his garden on Sunday?

Sunday: $3\frac{1}{2} - \frac{3}{4} = 2\frac{3}{4}$

$2\frac{3}{4}$ hours

Challenge

6. Complete each problem using the given digits. All numbers should be in simplest form.

(a) 1, 2, 3

$1\boxed{\frac{1}{6}} - \boxed{\frac{2}{3}} = \boxed{\frac{1}{2}}$

(b) 3, 4, 7, 8, 9, 21

$9\boxed{\frac{5}{21}} - \boxed{\frac{2}{3}} = 8\boxed{\frac{4}{7}}$

164 7-5 Subtracting a Fraction from a Mixed Number

Exercise 6 • pages 165–166

Exercise 6

Basics

1. Subtract.

(a) $7\frac{5}{16} - 2\frac{7}{8} = 5\frac{5}{16} - \frac{7}{8}$
$= 5\frac{5}{16} - \boxed{\frac{14}{16}}$
$= 4\boxed{\frac{21}{16}} - \boxed{\frac{14}{16}}$
$= 4\boxed{\frac{7}{16}}$

(b) $6\frac{1}{3} - 3\frac{7}{12} = 3\frac{1}{3} - \frac{7}{12}$
$= 3\boxed{\frac{4}{12}} - \frac{7}{12}$
$= 2\boxed{\frac{16}{12}} - \frac{7}{12}$
$= 2\boxed{\frac{9}{12}}$
$= 2\boxed{\frac{3}{4}}$

Practice

2. Subtract. Express each answer as a mixed number in simplest form.

(a) $8 - 4\frac{1}{5}$ $3\frac{4}{5}$

(b) $12 - 7\frac{5}{16}$ $4\frac{11}{16}$

(c) $6\frac{1}{2} - 2\frac{3}{10}$ $4\frac{1}{5}$

(d) $7\frac{5}{9} - 4\frac{1}{3}$ $3\frac{2}{9}$

(e) $9\frac{1}{6} - 3\frac{1}{3}$ $5\frac{5}{6}$

(f) $5\frac{1}{2} - 4\frac{5}{8}$ $\frac{7}{8}$

(g) $8\frac{1}{4} - 3\frac{7}{12}$ $4\frac{2}{3}$

(h) $10\frac{5}{12} - 3\frac{2}{3}$ $6\frac{3}{4}$

(i) $12 - 2\frac{5}{24} - 3\frac{1}{3}$ $6\frac{11}{24}$

(j) $14\frac{1}{2} - 6\frac{3}{4} - 1\frac{7}{8}$ $5\frac{7}{8}$

3. Eli had a board that was $9\frac{1}{2}$ m long. He cut off 2 pieces, one $3\frac{3}{10}$ m long and the other $2\frac{4}{5}$ m long. How long is the third piece of the board in meters?

$9\frac{1}{2} - 3\frac{3}{10} - 2\frac{4}{5} = 3\frac{2}{5}$

$3\frac{2}{5}$ meters

7-6 Subtracting Mixed Numbers   165

166   7-6 Subtracting Mixed Numbers

Exercise 7 • pages 167–170

Exercise 7

Check

Express answers 1 or greater as whole or mixed numbers. Use simplest form.

1 Find the values.

(a) $\frac{5}{9} + \frac{2}{3}$ $1\frac{2}{9}$

(b) $6\frac{9}{12} + \frac{5}{12}$ $7\frac{1}{6}$

(c) $10\frac{1}{3} - 7\frac{2}{3}$ $2\frac{2}{3}$

(d) $8\frac{4}{5} + 1\frac{7}{15}$ $10\frac{4}{15}$

(e) $3\frac{5}{14} - \frac{4}{7}$ $2\frac{11}{14}$

(f) $5\frac{13}{18} - 2\frac{5}{6}$ $2\frac{8}{9}$

(g) $8\frac{1}{6} + \frac{7}{30} + 3\frac{1}{2}$ $11\frac{27}{30}$

(h) $5\frac{27}{28} - 2\frac{3}{7} - \frac{13}{14}$ $2\frac{17}{28}$

2 Complete each pattern by following the rule. Use whole or mixed numbers when possible, and simplest form.

(a) Add $\frac{2}{3}$.

$\frac{2}{9}, \frac{8}{9}, \boxed{1\frac{5}{9}}, \boxed{2\frac{2}{9}}, \boxed{2\frac{8}{9}}, \boxed{3\frac{5}{9}}, \boxed{4\frac{2}{9}}, \boxed{4\frac{8}{9}}$

(b) Subtract $\frac{3}{8}$.

$3\frac{1}{2}, 3\frac{1}{8}, \boxed{2\frac{3}{4}}, \boxed{2\frac{3}{8}}, \boxed{2}, \boxed{1\frac{5}{8}}, \boxed{1\frac{1}{4}}, \boxed{\frac{7}{8}}$

(c) Add $2\frac{5}{12}$.

$\frac{1}{2}, \boxed{2\frac{11}{12}}, \boxed{5\frac{1}{3}}, \boxed{7\frac{3}{4}}, \boxed{10\frac{1}{6}}, \boxed{12\frac{7}{12}}, \boxed{15}, \boxed{17\frac{5}{12}}$

(d) Subtract $1\frac{3}{4}$.

$30\frac{1}{2}, \boxed{28\frac{3}{4}}, \boxed{27}, \boxed{25\frac{1}{4}}, \boxed{23\frac{1}{2}}, \boxed{21\frac{3}{4}}, \boxed{20}, \boxed{18\frac{1}{4}}$

3 Evan spent $2\frac{2}{3}$ hours practicing the piano in the morning. He spent another $1\frac{2}{3}$ hours practicing in the afternoon. How many hours did he practice in all?

$2\frac{2}{3} + 1\frac{2}{3} = 4\frac{1}{3}$

$4\frac{1}{3}$ hours

4 A railroad track-laying machine laid $1\frac{1}{5}$ km of track on Monday and $\frac{17}{20}$ km of track on Tuesday. How many more kilometers of track did it lay on Monday than on Tuesday?

$1\frac{1}{5} - \frac{17}{20} = \frac{7}{20}$

$\frac{7}{20}$ km

5 Package A weighs $4\frac{1}{2}$ pounds. Package B weighs $2\frac{1}{8}$ pounds more than Package A. What is the total weight of the two packages?

$4\frac{1}{2} + 2\frac{1}{8} + 4\frac{1}{2} = 11\frac{1}{8}$

$11\frac{1}{8}$ pounds

6 Catalina had $6\frac{3}{4}$ cups of flour. She used $2\frac{1}{2}$ cups of flour in one recipe and $2\frac{1}{4}$ cups of flour in another recipe. How much flour does she have left?

$6\frac{3}{4} - 2\frac{1}{2} - 2\frac{1}{4} = 2$

2 cups

Challenge

7 Write + or – in each ◯ to make each equation true.

(a) $\frac{13}{18} \oplus \frac{1}{9} \ominus \frac{1}{2} \oplus \frac{2}{3} = 1$

(b) $\frac{3}{4} \ominus \frac{13}{24} \oplus \frac{5}{8} \oplus \frac{1}{6} = 1$

8 This is a way to express 37 using five 3s and the addition symbol: $33 + 3 + \frac{3}{3}$.

(a) Use four 9s to express 100.

$99 + \frac{9}{9}$

(b) Use five 4s to express 55.

$44 + \frac{44}{4}$

(c) Use four 9s to express 20.

$9 + \frac{99}{9}$

(d) Use three each of 1, 3, 5, and 7 to express 20.

$1 + 3 + 5 + 7 + \frac{75}{75} + \frac{33}{11}$

Notes

Chapter 8 Multiplying a Fraction and a Whole Number — Overview

Suggested number of class periods: 9–10

	Lesson	Page	Resources	Objectives
	Chapter Opener	p. 243	TB: p. 181	Investigate multiplication as repeated addition of fractions.
1	Multiplying a Unit Fraction by a Whole Number	p. 244	TB: p. 182 WB: p. 171	Find the value of a whole number times a unit fraction.
2	Multiplying a Fraction by a Whole Number — Part 1	p. 246	TB: p. 185 WB: p. 174	Find the value of a whole number times a proper fraction.
3	Multiplying a Fraction by a Whole Number — Part 2	p. 248	TB: p. 188 WB: p. 176	Simplify the calculation before multiplying a fraction by a whole number.
4	Fraction of a Set	p. 250	TB: p. 192 WB: p. 179	Express the number of objects in part of a set as a fraction of the total number of objects in the set.
5	Multiplying a Whole Number by a Fraction — Part 1	p. 253	TB: p. 196 WB: p. 182	Find the fraction of a quantity. Multiply a fraction by a whole number.
6	Multiplying a Whole Number by a Fraction — Part 2	p. 257	TB: p. 201 WB: p. 185	Multiply a whole number by a fraction where the product is greater than 1.
7	Word Problems — Part 1	p. 260	TB: p. 204 WB: p. 188	Solve up to two-step word problems involving the fraction of a quantity.
8	Word Problems — Part 2	p. 263	TB: p. 207 WB: p. 192	Solve multi-step word problems involving fractions of a whole.
9	Practice	p. 266	TB: p. 210 WB: p. 196	Practice multiplying fractions.
	Workbook Solutions	p. 267		

Chapter 8 Multiplying a Fraction and a Whole Number

In this chapter, students deepen their understanding of fractions by learning how to multiply fractions and whole numbers.

Lessons 1–3 introduce multiplying a fraction by a whole number. Lesson 4 introduces the idea of fraction of a set, which is applied in Lessons 5–6 to multiply a whole number by a fraction.

Multiplying a Fraction by a Whole Number

Students will first look at examples where the fraction is the unit being multiplied. For example, $3 \times \frac{1}{5}$ means 3 groups of $\frac{1}{5}$. They can use repeated addition to solve this. They will start with unit fractions:

$3 \times \frac{1}{5} = \frac{1}{5} + \frac{1}{5} + \frac{1}{5}$

Ensure that students understand that we are multiplying units: 3×1 fifth = 3 fifths, or $\frac{3}{5}$. The unit (fifths) does not change, only the number of units changes.

Students will see that when multiplying a unit fraction by a whole number, the product of the numerator of the fraction and the whole number is the numerator of the answer:

$3 \times \frac{1}{5} = \frac{3 \times 1}{5} = \frac{3}{5}$

We can then apply this knowledge to non-unit fractions:

$3 \times \frac{2}{5}$

3×2 fifths = 6 fifths

Two methods will be covered to multiply a fraction by a whole number.

Method 1

Multiply the numerator of the fraction by the whole number:

$3 \times \frac{2}{5} = \frac{3 \times 2}{5} = \frac{6}{5} = 1\frac{1}{5}$

Method 2

Since the numerator counts the number of unit fractions, we can multiply it by the whole number first, then multiply the unit fraction by that product.

$3 \times \frac{2}{5}$

$3 \times 2 \times \frac{1}{5} = 6 \times \frac{1}{5}$

$= \frac{6}{5}$ or $1\frac{1}{5}$

In Lessons 1–2, students will simplify the product at the end.

$6 \times \frac{5}{8} = \frac{30}{8} = 3\frac{6}{8} = 3\frac{3}{4}$

In Lesson 3, they will learn that they can simplify the expression before finding the product:

$6 \times \frac{5}{8} = \frac{\overset{3}{\cancel{6}} \times 5}{\underset{4}{\cancel{8}}} = \frac{3 \times 5}{4} = \frac{15}{4} = 3\frac{3}{4}$

Fraction of a Set

In Lesson 4, students will formally learn how to find a fraction of a set.

This is a set of 12 toys:

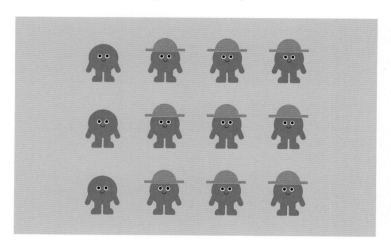

3 out of every 12 toys do not have hats. We can write this as a fraction. $\frac{3}{12}$ of the toys do not have hats.

$\frac{3}{12}$ can be simplified to $\frac{1}{4}$, so $\frac{1}{4}$ of the toys do not have hats. This can be illustrated by making 4 equal groups with the toys that do not have hats in 1 group.

Chapter 8 Multiplying a Fraction and a Whole Number   Notes

1 out of 4 groups is $\frac{1}{4}$.

Students may confuse the number of groups with the number of items in a group. Help students record their units (toys, groups of toys) so they can understand that: 12 toys is 4 groups of 3 toys.

Using the same example, we can also find the fraction of toys with hats. 9 out of 12, or $\frac{9}{12}$ of the toys have hats. That is 3 out of 4 groups of toys, or $\frac{3}{4}$ of the toys.

Students can check their work by thinking of parts and whole. In this case, they already found that $\frac{1}{4}$ of the toys, or 3 out of 12, do not have hats. 12 is the whole and 3 is a part. Remembering that $1 - \frac{1}{4} = \frac{3}{4}$, students could think about 12 (whole) − 3 (part) = 9 (other part), so $\frac{3}{4}$ of 12 = 9.

In Lesson 5, students will use this idea to multiply a whole number by a fraction. For example, in $\frac{1}{4} \times 16$, they are finding $\frac{1}{4}$ groups of 16, or $\frac{1}{4}$ times as much as 16.

There are 16 toys in the box. $\frac{1}{4}$ of them do not have hats. How many toys have hats?

Since we are told that $\frac{1}{4}$ of them do not have hats, we know to put them into 4 equal groups.

There are 4 equal groups. The number of toys in 1 group is 16 divided by 4, so we can find $\frac{1}{4} \times 16$ using division: $\frac{16}{4} = \frac{1}{4} \times 16$.

Then, $\frac{3}{4} \times 16$ is 3 times as many.

$\frac{1}{4} \times 16 = 4$

$\frac{3}{4} \times 16 = 3 \times 4 = 12$

We can see this easily in a bar model:

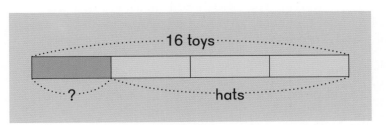

4 units → 16
1 unit → $\frac{16}{4}$ = 4   or   1 unit → $\frac{16}{4}$
3 units → 3 × 4 = 12   3 units → $3 \times \frac{16}{4}$ = 12

Chapter 8 Multiplying a Fraction and a Whole Number — Notes & Materials

Some problems lend themselves to thinking about multiplication of a fraction and a whole number as repeated addition. For example, if each scoop of batter measures $\frac{3}{4}$ cups, then 16 scoops of batter requires 12 cups of batter.

$16 \times \frac{3}{4} = \frac{48}{4} = 12$ cups

In other problems, students will find the fraction of a whole. For example, $\frac{3}{4}$ of a 16 mile hike is 12 miles.

$\frac{3}{4} \times 16 = 12$ miles

However, in both cases, the answer is the same. 16 groups of $\frac{3}{4}$ has the same answer as $\frac{3}{4}$ groups of 16.

Using either expression, we can simplify before calculating:

$\frac{3}{4} \times 16 = \frac{3 \times \cancel{16}^{4}}{\cancel{4}_{1}}$

Students will use bar models to understand and solve word problems with up to two steps involving multiplication of a fraction by a whole number.

Often we tell students that "of" tells us that we need to multiply, however, for some students this can be misleading. On the one hand, when we say $\frac{1}{6}$ of 12, we are looking for $\frac{1}{6} \times 12$, or $\frac{1}{6}$ as much as 12.

$\frac{1}{6}$ of 12 potatoes = $\frac{1}{6} \times 12$ potatoes = 2 potatoes

On the other hand, 6 out of 12 potatoes is $\frac{6}{12} = \frac{1}{2}$ of the set of potatoes — not 6 × 12, or 72 potatoes.

Remind students to study the context of the problem and model the problem with manipulatives or pictures rather than rely on keywords.

Materials

- 10-sided die, 3
- Bags, 3
- Beans or markers to cover bingo numbers
- Dry erase markers
- Fraction bars
- Fraction manipulatives
- Hula hoops
- Meter stick
- Paper circles, 6
- Recording sheet
- Slips of paper, 3 per student
- Small bean bags or other small, soft objects, 12
- Two-color counters
- Whiteboards
- Yarn

Blackline Masters

- Fraction Bingo Cards
- Fraction Bingo Game Board
- Multiplying Fractions Domino Cards
- Number Cards

Activities

Activities included in this chapter provide practice for multiplying fractions. They can be used after students complete the **Do** questions, or anytime additional practice is needed.

Chapter Opener

Objective

- Investigate multiplication as repeated addition of fractions.

Discuss the examples in the **Chapter Opener** and Alex's question. Emma suggests using repeated addition to find out how many of each ingredient it will take to triple the recipe.

This lesson may continue straight to Lesson 1.

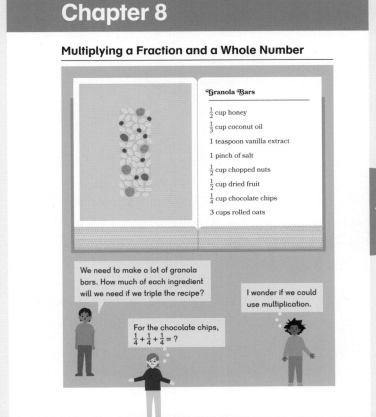

Lesson 1 Multiplying a Unit Fraction by a Whole Number

Objective
- Find the value of a whole number times a unit fraction.

Lesson Materials
- Fraction manipulatives

Think

Pose the **Think** problem and have students use the fraction tools or draw a model to solve the problem. Ask students to write an equation to show how they solved the problem.

Discuss student solutions. Students may have shown a repeated addition equation similar to Sofia's in **Learn**.

Learn

From their study of whole numbers, students know that multiplication can also be thought of as repeated addition. Ask students about how what they already learned relates to the equation and Sofia's thought.

Sofia and Emma share two ways of thinking about the problem. Sofia uses repeated addition. Just as 1 + 1 + 1 + 1 + 1 = 5, 1 eighth + 1 eighth + 1 eighth + 1 eighth + 1 eighth = 5 eighths.

Emma recalls that repeated addition is multiplication. Just as 5 × 1 = 5, 5 × 1 eighth = 5 eighths.

Discuss the equation that shows that the product of the numerator of the fraction and the whole number is the numerator of the answer. Since we are finding (5 times 1) eighths, we can write the multiplication as the numerator:

$\frac{5 \times 1}{8}$

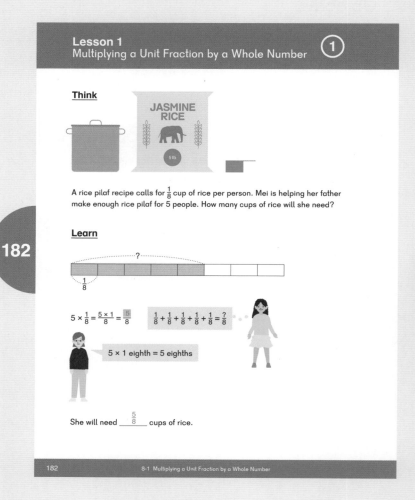

If necessary, discuss units other than fractions.

For example:

5 × 1 dog = 5 dogs

5 × 1 eighth = 5 eighths

Do

① (a) Alex helps students see that since $\frac{4}{5}$ is 4 units of $\frac{1}{5}$, 4 times $\frac{1}{5}$ is $\frac{4}{5}$.

In $\frac{4}{5}$, the numerator, 4, counts the number of fifths. Since we can use multiplication in repeated addition situations, we can write four times one-fifth, or 4 one-fifths.

(b) Mei sees that $\frac{5}{6}$ means 5 one-sixths or $5 \times \frac{1}{6}$, which is read "five times one-sixth."

⑤ Some students can just look at the bar model and know the total length of the blue ribbon is 1 m.

Total length:

1 unit ⟶ $\frac{1}{3}$ m
4 units ⟶ $4 \times \frac{1}{3}$ m = $\frac{4}{3}$ m

Others may solve by finding the length of the blue ribbon and adding that value to the length of the red ribbon:

$\frac{1}{3}$ m × 3 = 1 m

1 m + $\frac{1}{3}$ m = $1\frac{1}{3}$ m

Activity

▲ **Measuring**

Materials: Yarn, meter stick, recording sheet

Ask students how many cm are in 1 m. Then ask how many cm are in $\frac{1}{5}$ m.

Have each student cut a piece of string or yarn that is $\frac{1}{5}$ m long. Have them use the string to measure objects in the classroom that are $\frac{2}{5}$, $\frac{3}{5}$, and $\frac{4}{5}$ m long or wide, rounded to the nearest $\frac{1}{5}$ m.

After finding and measuring an objects, have students record their findings by writing a multiplication equation to express the length or width of the object.

Exercise 1 • page 171

Lesson 2 Multiplying a Fraction by a Whole Number — Part 1

Objective

- Find the value of a whole number times a proper fraction.

Lesson Materials

- Fraction bars

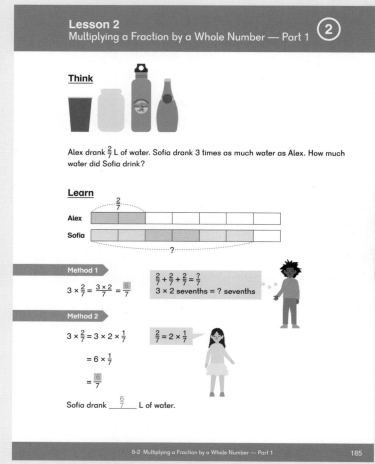

Think

Pose the **Think** problem and have students use the fraction bars to find the amount of water Sofia drank. Students can also draw a model or write an equation to solve the problem. Ask students how this problem is similar to and different from the **Think** problem in the previous lesson.

Discuss student solutions. Students may have chosen to write either addition or multiplication equations.

Learn

Have students discuss the comparison model and the two different methods.

Dion thinks of repeated addition for fractions and then recalls that it would be quicker to multiply.

Sofia thinks about unit fractions. She knows $\frac{2}{7} = 2 \times \frac{1}{7}$, so $3 \times \frac{2}{7} = 3 \times 2 \times \frac{1}{7}$.

Have students compare their methods from **Think** with the ones in the textbook.

Do

1 – 3 Ask students to use a different multiplication method to solve these problems. If necessary, remind them of Sofia's method on the previous page.

4 Have students identify which tick marks show eighths on the ruler (the ruler shows sixteenths). They could also look at a real ruler.

5 Have students share some of their solutions.

Activity

▲ Greatest Product

Materials: Three 10-sided dice

On each turn, players roll the dice and make a proper fraction with the numbers from two of the three dice. They multiply that fraction by the number on the third die.

Players can arrange the numbers in any order. The player with the greatest product is the winner.

Example:

Player may try the following:

$7 \times \frac{3}{9}$

$9 \times \frac{3}{7}$

$3 \times \frac{7}{9}$

$9 \times \frac{3}{7}$ results in the greatest product: $\frac{27}{7}$ or $3\frac{6}{7}$.

Exercise 2 • page 174

Do

1 Find the product of 4 and $\frac{2}{9}$.

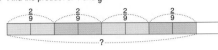

$4 \times \frac{2}{9} = \frac{4 \times 2}{9} = \frac{8}{9}$

2 Find the product of 5 and $\frac{2}{3}$.

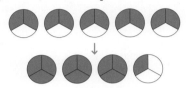

$5 \times \frac{2}{3} = \frac{5 \times 2}{3} = \frac{10}{3} = 3\frac{1}{3}$

3 Find the product of 4 and $\frac{3}{4}$.

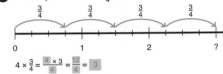

$4 \times \frac{3}{4} = \frac{4 \times 3}{4} = \frac{12}{4} = 3$

4 Mei is stapling 5 paper rectangles end-to-end on a bulletin board to make a border. Each rectangle is $\frac{7}{8}$ inches long. How long is the border?

$5 \times \frac{7}{8} = \frac{5 \times 7}{8} = \frac{35}{8} = 4\frac{3}{8}$

5 Multiply. Express each answer in its simplest form.

(a) $2 \times \frac{2}{5}$ $\frac{4}{5}$ (b) $2 \times \frac{3}{7}$ $\frac{6}{7}$

(c) $3 \times \frac{2}{9}$ $\frac{2}{3}$ (d) $5 \times \frac{3}{4}$ $3\frac{3}{4}$

(e) $2 \times \frac{4}{8}$ 1 (f) $3 \times \frac{5}{2}$ $7\frac{1}{2}$

(g) $6 \times \frac{3}{4}$ $4\frac{1}{2}$ (h) $3 \times \frac{4}{5}$ $2\frac{2}{5}$

6

A bag of broccoli weighs $\frac{3}{4}$ lb. Janice bought 3 bags of broccoli. How many pounds of broccoli did she buy?

$3 \times \frac{3}{4} = \frac{9}{4} = 2\frac{1}{4}$; $2\frac{1}{4}$ lbs

Exercise 2 • page 174

Lesson 3 Multiplying a Fraction by a Whole Number — Part 2

Objective

- Simplify the calculation before multiplying a fraction by a whole number.

Think

Pose the **Think** problem and have students draw a model or write an equation to solve the problem.

Ask students how this problem is similar to and different from the **Think** problem in the previous lesson. Discuss student solutions.

Learn

Discuss the methods in the textbook.

Method 1

In Method 1, the answer is calculated and then simplified.

Method 2

Instead of simplifying the answer, Mei simplifies the expression.

Students already know that the order of factors in a multiplication problem does not matter. Just as 2 × 3 has the same value as 3 × 2 (both equal 6), $\frac{2 \times 3}{8} = \frac{3 \times 2}{8}$.

Using his knowledge of equivalent fractions, Dion sees that 2 eighths is the same as 1 fourth. So $2 \times \frac{3}{8}$ can be simplified to $1 \times \frac{3}{4}$.

Students should think of the crossed off numerator and denominator as a short way of showing Dion's thinking. This idea will be further developed in **Do**.

Ask students why Alex is thinking of the factors of 2 and 8.

In Dimensions Math, students will be encouraged to simplify expressions before multiplying. This strategy is especially helpful when the numerator and/or denominator are large numbers. However, if a student chooses to simplify after multiplying, and is successful in showing the final answer in simplest form, that is acceptable.

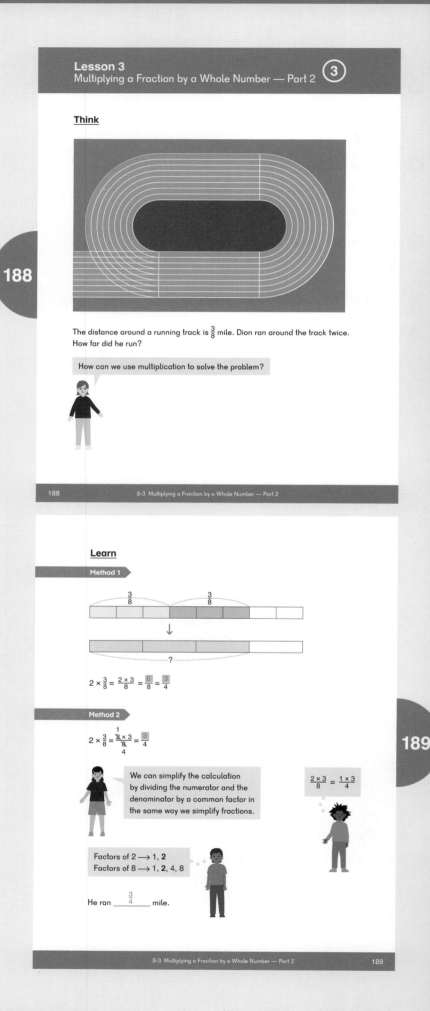

Teacher's Guide 4A Chapter 8

Do

For each question, discuss how the expression has been simplified before finding the final answer.

1. Sofia finds the factors of 3 and 9 to simplify the expression. Dion shows the shortcut method.

3. Emma finds common factors to simplify the expression.

4. Alex and Mei see that both 2 and 4 are factors of 12. Mei chooses to simplify the expression in two steps by dividing both the numerator and denominator by 2 twice. This strategy requires more steps, but is helpful when students are not easily identifying factors.

Another example of simplifying a fraction in two steps:

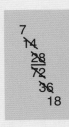

First, if the numerator and denominator are both even, they can be divided by the factor of 2. If this also results in an even numerator and denominator, they can again be divided by 2.

Activity

★ **How Many?**

Materials: Number Cards (BLM) 1–9

Have students use the numbers 1–9 once to fill the spaces and find valid equations.

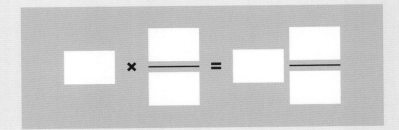

Sample solutions:

$2 \times \frac{7}{8} = 1\frac{3}{4}$ or $6 \times \frac{3}{8} = 2\frac{1}{4}$

Exercise 3 • page 176

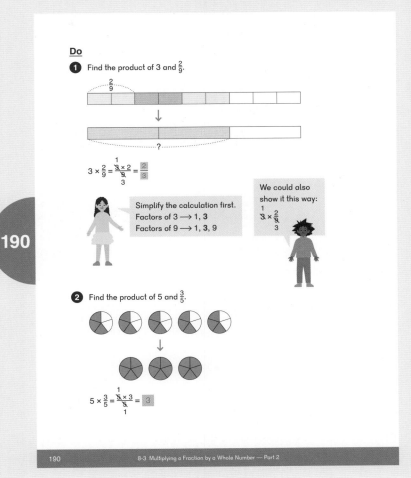

Lesson 4 Fraction of a Set

Objective

- Express the number of objects in part of a set as a fraction of the total number of objects in the set.

Lesson Materials

- Two-color counters

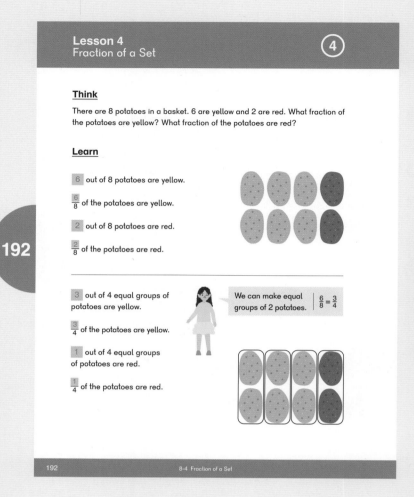

Think

Provide students with two-color counters to represent the potatoes and pose the **Think** problems.

Discuss student solutions.

Learn

Have students compare their solutions from **Think** with the ones in the textbook. Students have previously encountered fractions that are a part of 1 whole, such as 1 meter or 1 cake. Now the fraction is a part of 1 whole set, in this case, 1 set of potatoes.

When learning fractions of a set, students must always first identify what the set is, and how many objects are in the set. In this case, the set is 8 potatoes.

Most students will see that 6 out of 8 potatoes are yellow, so $\frac{6}{8}$ of the potatoes are yellow and $\frac{2}{8}$ of the potatoes are red.

Discuss Sofia's comment. $\frac{6}{8}$ simplifies to $\frac{3}{4}$. This means we can say that $\frac{3}{4}$ of the potatoes are yellow. Discuss the picture below Sofia where the potatoes are shown in groups. This helps students see that if we put the potatoes into equal groups, with only one color in a group, then we can say that 3 groups out of the 4 groups of potatoes are yellow. Similarly, one group out of the 4 groups of potatoes is red.

Do

1 (a) There are 5 animals in the set. 2 of the animals are owls so $\frac{2}{5}$ of the set are owls.

Ask students how problem (b) is similar to and different from problem (a). They should see that the total number of animals changes, but when the animals are put into equal groups with only one kind of animal in each group, we can see that 2 out of 5 groups are owls, so $\frac{2}{5}$ of the animals are owls.

Alex reminds us that $\frac{6}{15} = \frac{2}{5}$.

If $\frac{6}{15}$ of the animals are owls, then $\frac{15}{15} - \frac{6}{15}$, or $\frac{9}{15}$, of the animals are bats:

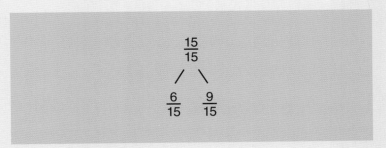

$\frac{9}{15}$ can be simplified to $\frac{3}{5}$.

It is important to note that in both (a) and (b), $\frac{2}{5}$ of the animals are owls. From the fraction alone, we cannot tell the total number of owls. We need to know the total in the set.

2 (b) The set consists of 12 counters in 4 equal groups. 3 out of 4 equal groups of counters are red.

(c) The set consists of 8 counters. 2 counters are red. Students can find 2 out of 8 of the counters are red, or group the counters to see 1 group of 4 is red.

3 Students should give their answers in simplest form.

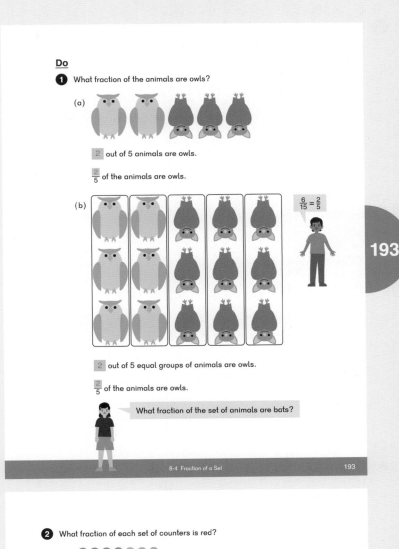

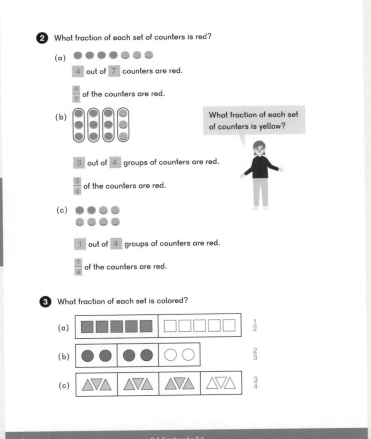

4 Dion relates the number of groups to simplest form.

$\frac{6}{8} = \frac{3}{4}$

Ask students to imagine that the columns of toasters are circled, and are therefore each a group. They should see that 3 out of a total of 4 groups is $\frac{3}{4}$, so when we group equal parts in a set, we are simplifying the fraction.

5 Students should see that the problem is asking about 25 out of 100. They can see that 25 out of 100 is $\frac{25}{100}$, which is $\frac{1}{4}$ in simplest form.

Students can first simplify $\frac{25}{100}$ to find the fraction of mandarin oranges ($\frac{1}{4}$), then subtract $\frac{1}{4}$ from 1 to find the fraction of navel oranges. Or, they can subtract 25 from 100 to find the total number of navel oranges (75) and then simplify the fraction $\frac{75}{100}$ to $\frac{1}{4}$.

Activity

▲ Fraction Toss

Materials: Hula hoops, 12 small bean bags or other small, soft objects

On each round, students take turns attempting to toss all 12 of the bean bags into the hula hoop. After each turn, they say what fraction of the bean bags landed inside the hula hoop. The numerator of the simplest form of the fraction is the player's score.

For example, Player One lands $\frac{8}{12}$, or $\frac{2}{3}$, of the bean bags in the hoop. His score is 2.

Player Two lands $\frac{6}{12}$ of the bean bags in the hoop. The simplest form of $\frac{6}{12}$ is $\frac{1}{2}$. His score is 1.

Player Three lands $\frac{7}{12}$ of the bean bags in the hoop. Her score is 7.

Player Four lands all twelve of the beans bags in the hoop. His score for $\frac{12}{12}$ is a whole number, 1. He has no numerator and scores 0 points on this round.

The highest score in each round is the winner.

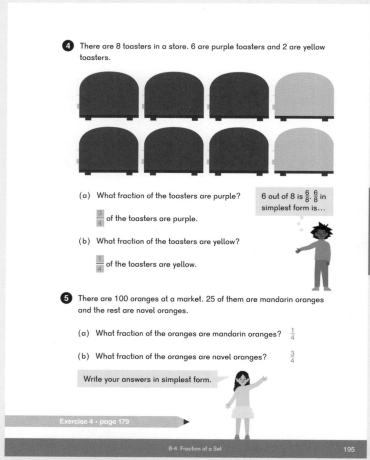

Substitute paper plates and tiddly winks or any flat counters for indoor practice.

Exercise 4 • page 179

Lesson 5 Multiplying a Whole Number by a Fraction — Part 1

Objectives

- Find the fraction of a quantity.
- Multiply a fraction by a whole number.

Lesson Materials

- Two-color counters

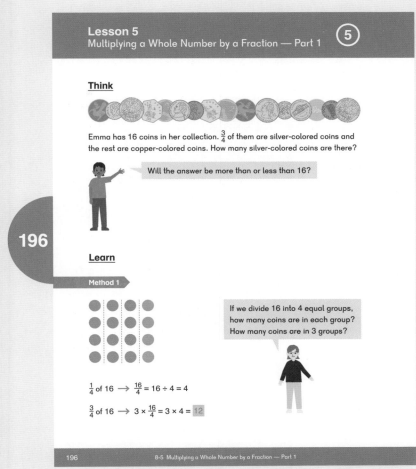

Think

Provide students with two-color counters to represent Emma's coins and pose the **Think** problem. Ask students how this problem is similar to and different from the **Think** problem in the previous lesson. They should see that this **Think** problem is looking for a quantity, while the problem in the previous lesson was looking for a fraction.

Ask students to think of a way to solve the problem before looking at the methods provided in the textbook, based on what they learned in the previous lesson.

Alex asks students to think about how many silver-colored coins there will be. If there are 16 coins and part of them are silver-colored, there must be fewer than 16 silver-colored coins.

Learn

Have students discuss the **Learn** examples.

Method 1

We need to find $\frac{3}{4}$ of 16. We know from the denominator that there are 4 equal groups. Emma tells us that if we make 4 equal groups, we can find the number of coins in 3 of the groups that are silver-colored. Since 16 divided by 4 is 4, each group has 4 coins. Three such groups will have 3 × 4, or 12 coins.

Method 2

We can draw a bar model. First, the whole set consists of 16 coins.

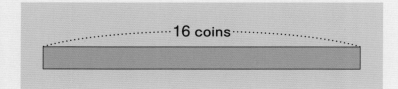

Mei explains that $\frac{3}{4}$ of 16 is $\frac{3}{4}$ times 16. Since we have fourths, we divide the bar into 4 equal parts.

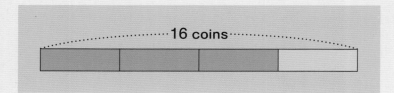

Each unit is 1 fourth of the whole:

4 units ⟶ 16
1 unit ⟶ $\frac{16}{4}$ = 4

There are 4 copper-colored coins.

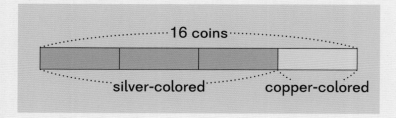

From here, there are two paths to solve the problem.

Path 1:

3 units ⟶ 3 × $\frac{16}{4}$ = 3 × 4 = 12

There are 12 silver-colored coins.

Path 2:

We know the whole is 16 coins and $\frac{1}{4}$ of the whole is 4 coins. We can subtract 1 unit (or 4 coins) to solve for the 3 units of silver-colored coins:
16 − 4 = 12

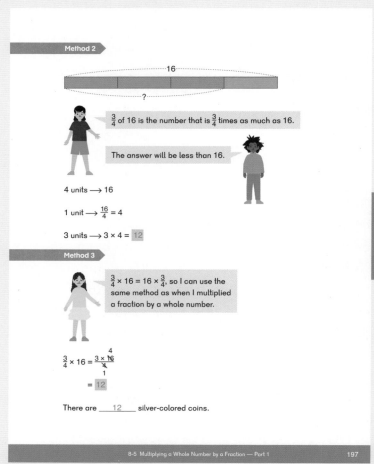

Method 3

Sofia knows that since the answer will be the same for $\frac{3}{4}$ × 16 as for 16 × $\frac{3}{4}$, we can solve the problem in the same way as we did when finding 16 groups of $\frac{3}{4}$, and multiply the numerator by the whole number.

Do

1 (a) Divide 12 into 3 equal groups to find how many are in each group, since each group is $\frac{1}{3}$ of 12.

(b) $\frac{1}{3} \times 12 = \frac{12}{3}$

$\frac{2}{3} = 2 \times \frac{1}{3}$, so $\frac{2}{3}$ of $12 = 2 \times \frac{12}{3} = 2 \times 4$.

Students can simplify the expression $\frac{2 \times \cancel{12}^4}{\cancel{3}_1} = 8$.

2 (a) Students can simplify the expression $\frac{1 \times \cancel{18}^3}{\cancel{6}_1} = 3$.

(b) If $\frac{1}{6}$ of 18 is 3, then 5 one-sixths of 18 is 5×3, or 15.

3 — **4** Possible questions to ask students:

- "What fraction do the units represent?"
- "What is the value of the whole?"
- "How can we find the value of 1 unit?"
- "Once we find the value of 1 unit, are we finished?"

3 Since we need to find 3 fourths, the bar model shows the whole, 24, divided into 4 equal units. Two methods to solve are then shown:

(a) Find the value of 1 unit first by dividing 24 by 4, and simplifying $\frac{24}{4}$ to 6. Multiply to find 3 units of 6, which is 18.

(b) Write a multiplication equation. Mei simplifies before she multiplies.

4 Since we need to find 4 fifths, the bar model shows the whole, 15, divided into 5 equal units. Two methods to solve, similar to **3**, are then shown:

(a) Find the value of 1 unit first by dividing 15 by 5, and simplifying $\frac{15}{5}$ to 3. Multiply to find 4 units of 3, which is 12.

(b) Write a multiplication equation and simplify.

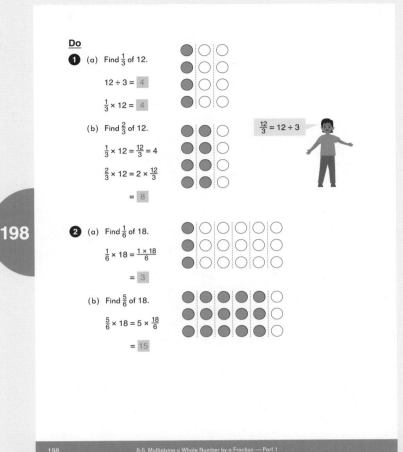

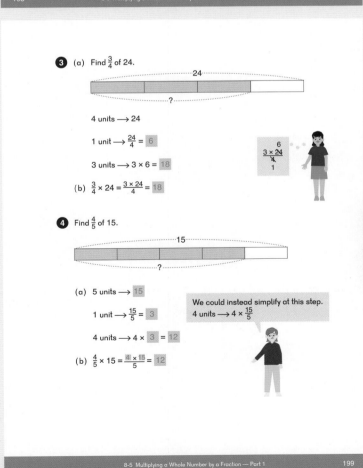

Activity

▲ Fraction Bingo 24

Materials: Fraction Bingo Cards (BLM), Fraction Bingo Game Board (BLM), beans or markers to cover bingo numbers

Have students fill in their game boards with the numbers listed. They may choose what numbers to include.

To call a number, turn over a Fraction Bingo Card. Students multiply the fraction on the card by 24 and cover the product on their game boards.

For example, if the Caller says:

- $\frac{1}{3} \times 24$: Students find $\frac{24}{3} = 8$ and cover 8 on their game boards.
- $\frac{4}{6} \times 24$: Students find $\frac{4}{6} = \frac{2}{3}$ and $\frac{2}{3}$ of 24 = 16. They cover 16 on their game boards.

The first person to get 4 in a row wins.

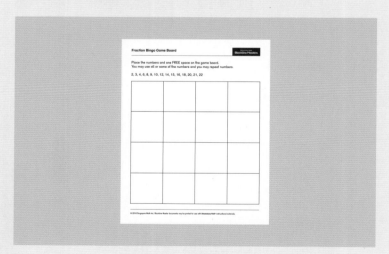

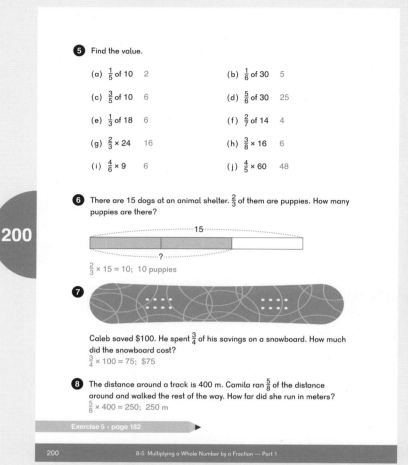

Exercise 5 • page 182

Lesson 6 Multiplying a Whole Number by a Fraction — Part 2

Objective
- Multiply a whole number by a fraction where the product is greater than 1.

Lesson Materials
- 6 paper circles

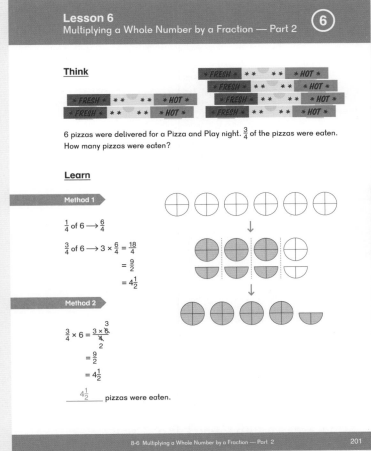

Think

Provide students with 6 paper circles to represent the pizzas in the **Think** problem. Tell students they can fold, cut, or color the paper circles to show how many pizzas were eaten.

Ensure that students understand the answer to the question, "How many pizzas were eaten?" does not have to be a whole number.

Discuss student solutions.

Learn

All 6 of the pizzas are divided into fourths and then the fourths are put into four equal groups.

Students can calculate $\frac{3}{4} \times 6$ using one of two different methods.

Method 1

Students find $\frac{1}{4}$ of 6, then multiply the answer by 3. Note that after the first step, $\frac{1}{4}$ of 6, they should not simplify the improper fraction to a mixed number, and instead leave it as an improper fraction. Students do not formally multiply mixed numbers until Dimensions Math 5.

In the previous lesson, students could simplify the value for the unit fraction times the whole number because the answer was a whole number.

$\frac{18}{4}$ can be simplified to $\frac{9}{2}$ or $4\frac{1}{2}$.

Method 2

Students multiply $\frac{3}{4}$ times 6.

Students may also have simply placed 4 of the 6 whole pizzas in groups and divided the 2 remaining pizzas into fourths. They then put 2 fourths in each group to find there are $1\frac{1}{2}$ pizzas in each group. They find how many pizzas are in 3 groups:

$1\frac{1}{2} + 1\frac{1}{2} + 1\frac{1}{2} = 1 + 1 + 1 + \frac{1}{2} + \frac{1}{2} + \frac{1}{2}$, or $4\frac{1}{2}$ pizzas.

Have students compare their methods from **Think** with the ones in the textbook.

Do

1–**4** For each question, discuss the different calculations shown.

1 (b) Students can use the improper fraction, $\frac{10}{3}$, to multiply by 2, rather than the mixed number $3\frac{1}{3}$.

There are 30 pieces. If $\frac{1}{3}$ of 30 is 10, then $\frac{2}{3}$ is twice as many as 10. There are 20 thirds.

$\frac{20}{3} = 6\frac{2}{3}$

Although this lesson does not teach multiplication of a mixed number by a whole number, some students may see that because $\frac{1}{3}$ of 10 is $3\frac{1}{3}$, $\frac{2}{3}$ of 10 must be $2 \times 3\frac{1}{3} = 6\frac{2}{3}$.

2 Ask students:

- "What is the whole?" (4 m)
- "What fraction do the units on the bar model represent?" ($\frac{1}{5}$)
- "How can we find the value of 1 unit?" (1 unit ⟶ 4 m divided by 5, or $\frac{4}{5}$ m)
- "How can we use that knowledge to find $\frac{3}{5}$ of 4?" (3 units ⟶ $3 \times \frac{4}{5}$ m)

④ Dion simplifies before multiplying.

⑤ Have students share their methods for solving some of the problems.

⑥ Ask students:

- "What is the whole?" (8 ft)
- "What fraction do the units on the bar model represent?" ($\frac{1}{3}$)
- "How can we find the value of 1 unit?" (1 unit ⟶ 8 ft divided by 3 or $\frac{8}{3}$ ft)
- "How can we find $\frac{2}{3}$ of 8 ft?" (2 units ⟶ 2 × $\frac{8}{3}$ ft)

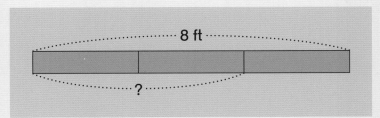

3 units ⟶ 8 ft
1 unit ⟶ $\frac{8}{3}$ ft
2 units ⟶ 2 × $\frac{8}{3}$ ft = $\frac{16}{3}$ ft
The piece he cut off is $5\frac{1}{3}$ ft.

Activity

▲ **Paper Bag Word Problems**

Materials: 3 bags, 3 slips of paper filled out by each student with the following:

- Bag #1: whole number
- Bag #2: fraction
- Bag #3: noun

After each student fills out three slips with a whole number, fraction, and noun, and places them into each of the respective bags, shake the bags and have students draw 1 slip of paper from each bag.

Students write word problems requiring multiplication based on the numbers and nouns on their slips of paper. Once written, students can trade word problems with a partner and solve.

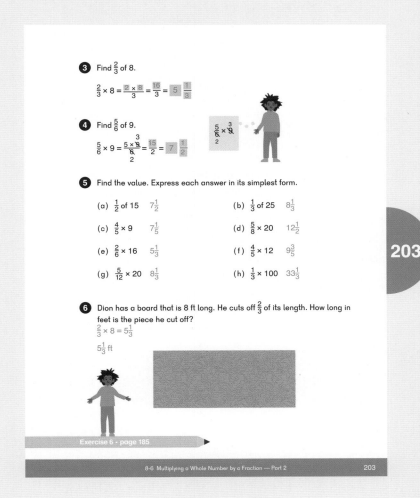

For example, for 4, $\frac{1}{2}$, and dog, students could write: If 1 dog eats $\frac{1}{2}$ cup of dog food, how much dog food do 4 dogs eat?

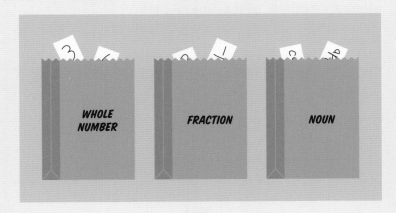

Exercise 6 • page 185

Lesson 7 Word Problems — Part 1

Objective

- Solve up to two-step word problems involving the fraction of a quantity.

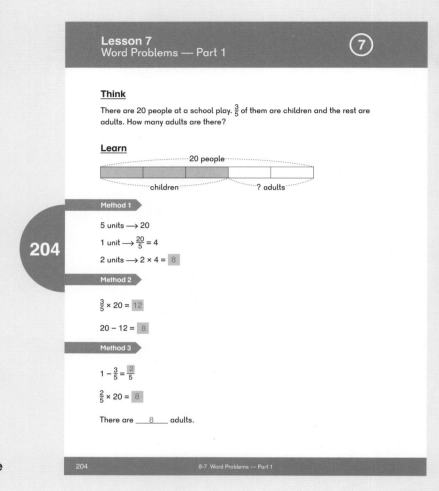

Think

Pose the **Think** problem and have students write an equation to show their solutions. They may also choose to draw a model. If students finish quickly, challenge them to find other methods to solve the problem.

Compare and discuss student solutions.

Learn

Have students compare the three methods shown in **Learn** with their solutions from **Think**.

Method 1

Ensure students see that the model is 5 units because we are asked to find fifths of the whole (20 people).

Method 2

Find the number of children first by multiplying $\frac{3}{5} \times 20$, then subtracting the product from the total to find the number of adults at the school play.

Method 3

Some students may look at the model and see that the number of adults is $\frac{2}{5}$ of the people, then find $\frac{2}{5}$ of the whole.

Do

1 – **2** In these problems, a value of a fractional part, instead of a whole, is given. Discuss the given models.

1 In this problem, the value of a fractional part is given, and the whole must be found.

Ask students:

- "What is alike and different between **1** and the **Think** problem?"
- "Are we given a part or a whole?" (part)
- "Are we finding a part or a whole?" (whole)
- "What fraction do the units represent?" ($\frac{1}{5}$)
- "How can we find the value of 1 unit?" (1 unit ⟶ $\frac{12}{2} = 6$)
- "How can we find the whole, or $\frac{5}{5}$?" (5 units ⟶ 5 × 6 = 30)

2 In this problem, a value for the fractional part, $\frac{3}{5}$, is given and students must find the value of the other part, $\frac{2}{5}$.

Ask students:

- "What is alike and different between **2** and the **Think** problem?"
- "Are we given a part or a whole?" (part)
- "Are we finding a part or a whole?" (part)
- "What fraction does each unit represent?" ($\frac{1}{5}$)
- "How can we find the value of 1 unit?" (1 unit ⟶ $\frac{36}{3} = 12$)
- "How can we find the part, or $\frac{2}{5}$?" (2 units ⟶ 2 × 12 = 24)

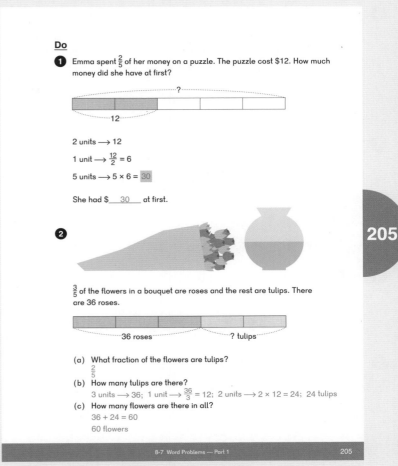

3 Ask students:

- "In what units are the parts given?" (eighths)
- "Are we finding a part or a whole?" (part)
- "How many eighths is the part we are finding?" (5 eighths)

4 Ask students:

- "In what units are the parts given?" (thirds)
- "Are we finding a part or a whole?" (whole)
- "We know the value of how many thirds?" (2 thirds)
- "How many thirds do we need to find?" (3 thirds)

5–**7** Have students draw a model to help them solve the problem. Continue to ask questions similar to those above if needed.

5
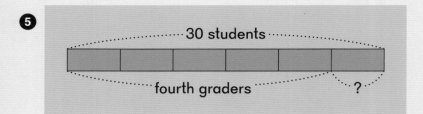

6
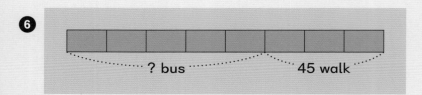

7
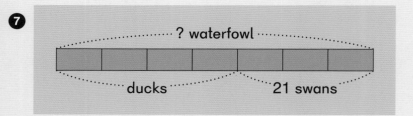

Exercise 7 • page 188

---

3 Dion had \$40. He used $\frac{3}{8}$ of it to buy a game. How much money did he have left?

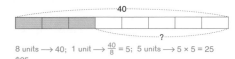

8 units ⟶ 40; 1 unit ⟶ $\frac{40}{8}$ = 5; 5 units ⟶ 5 × 5 = 25
\$25

4 In a coin collection, $\frac{1}{3}$ of the coins are silver coins and the rest are gold coins. If there are 14 gold coins, how many coins are there altogether?

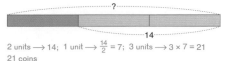

2 units ⟶ 14; 1 unit ⟶ $\frac{14}{2}$ = 7; 3 units ⟶ 3 × 7 = 21
21 coins

5 There are 30 fourth and fifth graders on a basketball team. $\frac{5}{6}$ of them are fourth graders. How many fifth graders are on the team?
$\frac{1}{6}$ × 30 = 5; 5 fifth graders

6 $\frac{5}{8}$ of the fourth graders ride the bus to school and the rest walk to school. If 45 fourth graders walk to school, how many ride the bus to school?
3 units ⟶ 45; 1 unit ⟶ $\frac{45}{3}$ = 15
5 units ⟶ 5 × 15 = 75; 75 fourth graders

7 $\frac{4}{7}$ of the waterfowl in a pond are ducks and the rest are swans. There are 21 swans. How many waterfowl are in the pond?
3 units ⟶ 21
1 unit ⟶ $\frac{21}{3}$ = 7
7 units ⟶ 7 × 7 = 49
49 waterfowl

Exercise 7 • page 188

Lesson 8 Word Problems — Part 2

Objective

- Solve multi-step word problems involving fractions of a whole.

Think

Pose the **Think** problem. If students finish quickly, challenge them to find another method to solve the problem. Focus student attention on the model. Ask them how this relates to what they learned before about a fraction of a set.

Compare and discuss student solutions.

Learn

Have students compare the methods shown in **Learn** with their solutions from **Think**.

Mei subtracts the number of green envelopes from the total to get the number of yellow envelopes. She writes the fraction of yellow envelopes ($\frac{8}{24}$) and simplifies.

Alex finds the fraction of green envelopes and simplifies to $\frac{2}{3}$. He subtracts that from the whole of 1 to find that $\frac{1}{3}$ of the envelopes are yellow.

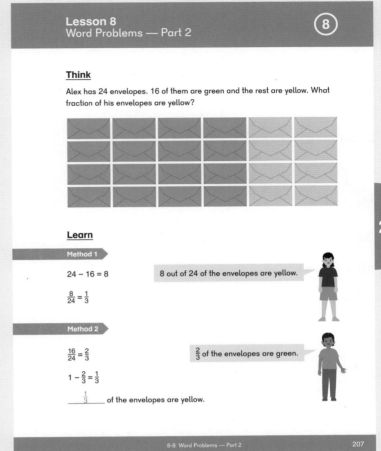

Do

2 **Method 1**

Fraction of children wearing hats:

$\frac{25}{40} = \frac{5}{8}$

Fraction of children not wearing hats:

$1 - \frac{5}{8} = \frac{3}{8}$

Method 2

Number of children not wearing hats:

$40 - 25 = 15$

Fraction of children not wearing hats:

$\frac{15}{40} = \frac{3}{8}$

4 We can represent each fruit as a fraction of the whole set:

Peaches: $\frac{8}{48}$

Apples: $\frac{16}{48}$

Oranges: $48 - 16 - 8 = 24$, so $\frac{24}{48}$

Each of these fractions can then be simplified.

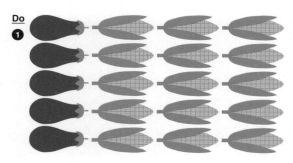

5. Discuss the bar model and how it relates to the problem.

6.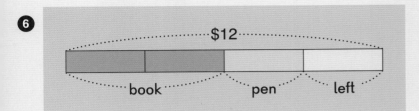

Emma finds $\frac{1}{2}$ of 12 and $\frac{1}{4}$ of 12. She can add 6 + 3 to find the money spent, and subtract that sum from the total to find the amount of money left.

Sofia uses the sum of $\frac{1}{2}$ and $\frac{1}{4}$ ($\frac{3}{4}$). She subtracts $\frac{3}{4}$ from 1, then multiplies $\frac{1}{4} \times 12$ to get her answer.

7. Students can draw different bar models to help them solve the problem. Example of a bar model that includes all information in the problem:

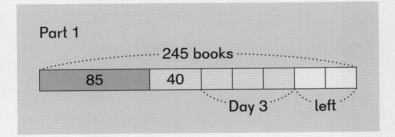

Students can first subtract to find the number of books left after the second day:

245 − 85 − 40 = 120

If needed, students can draw a second bar model to find the answers to (a) and (b):

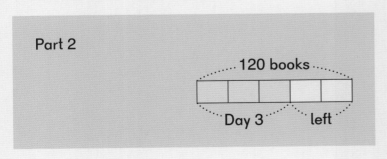

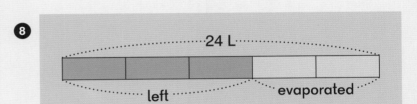

5 units → 24 L
1 unit → $\frac{24}{5}$ L
3 units → 3 × $\frac{24}{5}$ L = $14\frac{2}{5}$ L
$14\frac{2}{5}$ L of water are left in the tub.

Exercise 8 • page 192

Lesson 9 Practice

Objective

- Practice multiplying fractions.

After students complete the **Practice** in the textbook, have them continue to practice multiplying fractions and whole numbers with activities from the chapter.

Exercise 9 • page 196

Brain Works

★ Multiplying Fractions Dominos

Materials: Multiplying Fractions Domino Cards (BLM)

Begin with the domino card labeled "Start." Players match a multiplication expression on one half of the domino card with the correct solution on one half of another domino card.

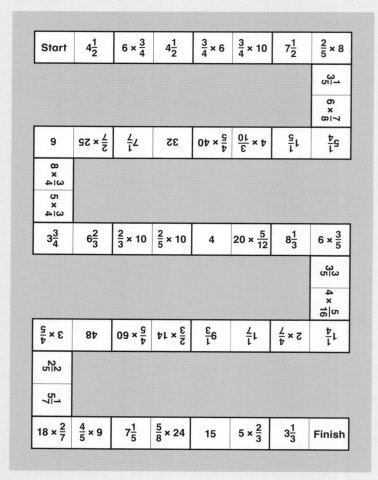

Lesson 9 Practice — P 9

① Multiply. Express each answer in its simplest form.

(a) $2 \times \frac{1}{3}$ $\frac{2}{3}$ (b) $\frac{2}{7} \times 3$ $\frac{6}{7}$ (c) $6 \times \frac{1}{8}$ $\frac{3}{4}$

(d) $\frac{4}{9} \times 15$ $6\frac{2}{3}$ (e) $\frac{2}{5} \times 5$ 2 (f) $8 \times \frac{3}{4}$ 6

② A bread recipe calls for $\frac{3}{4}$ cups of flour to make one loaf of bread. Chad wants to make 3 loaves of bread. How much flour should he use?
$\frac{3}{4} \times 3 = 2\frac{1}{4}$; $2\frac{1}{4}$ cups

③ There are 28 students in a fourth grade class. 16 of them are wearing sneakers. What fraction of the students are not wearing sneakers?
$28 - 16 = 12$; $\frac{12}{28} = \frac{3}{7}$; $\frac{3}{7}$ of the students

④ Johnathan earned some money gardening. He spent $\frac{1}{4}$ of the money he earned on a board game that cost $11. How much money did he earn gardening?
1 unit ⟶ 11
4 units ⟶ 4 × 11 = 44; $44

⑤ Ashna had 8 ft of ribbon. She used $\frac{3}{5}$ of it to wrap presents. How many feet of ribbon does she have left?
$1 - \frac{3}{5} = \frac{2}{5}$; $\frac{2}{5} \times 8 = 3\frac{1}{5}$; $3\frac{1}{5}$ ft

⑥ $\frac{2}{5}$ of the students in the Language Club are learning Spanish, $\frac{3}{10}$ of them are learning Mandarin, and the rest are learning French. If 30 students are learning Mandarin, how many students are in the club?
3 units ⟶ 30; 1 unit ⟶ $\frac{30}{3} = 10$; 10 units ⟶ 10 × 10 = 100
100 students

Exercise 9 • page 196

Exercise 1 • pages 171–173

Chapter 8 Multiplying a Fraction and a Whole Number

Exercise 1

Basics

1. Find the value of 3 groups of $\frac{1}{9}$.

    $3 \times \frac{1}{9} = \frac{3 \times 1}{9}$

    $= \boxed{\frac{3}{9}}$

    $= \boxed{\frac{1}{3}}$

2. Find the value of 5 groups of $\frac{1}{3}$.

    $5 \times \frac{1}{3} = \frac{5 \times 1}{3}$

    $= \boxed{\frac{5}{3}}$

    $= 1\boxed{\frac{2}{3}}$

    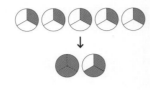

3. How many $\frac{1}{4}$s are in $5\frac{3}{4}$?

    $5\frac{3}{4} = \boxed{\frac{23}{4}}$

    $= \boxed{23} \times \frac{1}{4}$

    There are ___23___ $\frac{1}{4}$s in $5\frac{3}{4}$.

4. Express the products as mixed numbers in simplest form.

    (a) Find the product of 16 and $\frac{1}{5}$.

    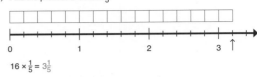

    $16 \times \frac{1}{5} = 3\frac{1}{5}$

    (b) Find the product of 16 and $\frac{1}{6}$.

    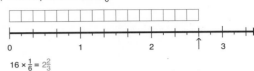

    $16 \times \frac{1}{6} = 2\frac{2}{3}$

Practice

Express answers 1 or greater as whole or mixed numbers. Use simplest form.

5. Multiply.

    (a) $5 \times \frac{1}{8}$ $\frac{5}{8}$
    (b) $7 \times \frac{1}{12}$ $\frac{7}{12}$
    (c) $5 \times \frac{1}{5}$ 1
    (d) $8 \times \frac{1}{10}$ $\frac{4}{5}$
    (e) $6 \times \frac{1}{3}$ 2
    (f) $20 \times \frac{1}{4}$ 5
    (g) $8 \times \frac{1}{5}$ $1\frac{3}{5}$
    (h) $10 \times \frac{1}{8}$ $1\frac{1}{4}$
    (i) $15 \times \frac{1}{9}$ $1\frac{2}{3}$
    (j) $30 \times \frac{1}{15}$ 2

6. A jug has 4 L of juice. $\frac{1}{3}$ L of juice was poured into each of 10 glasses. How many liters of juice are still in the jug?

    $10 \times \frac{1}{3} = 3\frac{1}{3}$

    $4 - 3\frac{1}{3} = \frac{2}{3}$

    $\frac{2}{3}$ L

Exercise 2 • pages 174–175

Exercise 2

Basics

1 Find the value of 3 groups of $\frac{3}{4}$.

$3 \times \frac{3}{4} = \frac{3 \times 3}{4}$

$= \boxed{\frac{9}{4}}$

$= 2\boxed{\frac{1}{4}}$

2 Find the product of 5 and $\frac{2}{3}$.

$5 \times \frac{2}{3} = \frac{5 \times 2}{3} = \boxed{\frac{10}{3}} = 3\boxed{\frac{1}{3}}$

3 Find the product of 7 and $\frac{2}{9}$. Express the answer as a mixed number.

$7 \times \frac{2}{9} = \frac{14}{9} = 1\frac{5}{9}$

Practice

Express answers 1 or greater as whole or mixed numbers. Use simplest form.

4 Multiply.

(a) $2 \times \frac{3}{7}$ $\frac{6}{7}$

(b) $7 \times \frac{5}{12}$ $2\frac{11}{12}$

(c) $3 \times \frac{3}{5}$ $1\frac{4}{5}$

(d) $9 \times \frac{3}{10}$ $2\frac{7}{10}$

(e) $5 \times \frac{5}{6}$ $4\frac{1}{6}$

(f) $5 \times \frac{3}{7}$ $2\frac{1}{7}$

(g) $7 \times \frac{5}{11}$ $3\frac{2}{11}$

(h) $8 \times \frac{7}{15}$ $3\frac{11}{15}$

5 A park $1\frac{3}{5}$ miles from Carter's home has a trail around it that is $\frac{9}{10}$ mile long. Carter ran to the park, ran the trail 3 times, and then ran home. How far did he run in all?

$3 \times \frac{9}{10} = 2\frac{7}{10}$

$1\frac{3}{5} + 2\frac{7}{10} + 1\frac{3}{5} = 5\frac{9}{10}$

$5\frac{9}{10}$ miles

Exercise 3 • pages 176–178

Exercise 3

Basics

1 Find the value of 4 groups of $\frac{3}{4}$.

(a) $4 \times \frac{3}{4} = \frac{4 \times 3}{4}$

$= \boxed{\frac{12}{4}}$

$= \boxed{3}$

(b) $4 \times \frac{3}{4} = \frac{\cancel{4} \times 3}{\cancel{4}_1} = \boxed{3}$

2 Find the value of 4 groups of $\frac{7}{10}$.

$4 \times \frac{7}{10} = \frac{\cancel{4}^2 \times 7}{\cancel{10}_5}$

$= \boxed{\frac{14}{5}}$

$= 2\boxed{\frac{4}{5}}$

3 Find the product of 6 and $\frac{5}{9}$.

$6 \times \frac{5}{9} = \frac{6 \times 5}{9} = \boxed{\frac{10}{3}} = 3\boxed{\frac{1}{3}}$

Practice

Express answers 1 or greater as whole or mixed numbers. Use simplest form.

4 Multiply.

(a) $3 \times \frac{2}{9}$ $\frac{2}{3}$

(b) $8 \times \frac{3}{4}$ 6

(c) $15 \times \frac{2}{5}$ 6

(d) $3 \times \frac{5}{6}$ $2\frac{1}{2}$

(e) $8 \times \frac{3}{10}$ $2\frac{2}{5}$

(f) $10 \times \frac{3}{8}$ $3\frac{3}{4}$

(g) $15 \times \frac{4}{9}$ $6\frac{2}{3}$

(h) $15 \times \frac{5}{12}$ $6\frac{1}{4}$

5 One wall of Allysa's room is 3 m long. She is making a ceiling border for that wall by pasting $\frac{3}{10}$ m pieces of wallpaper trim end-to-end. So far she has pasted 9 pieces of trim. How long does she have to make the last piece of trim to complete the project?

$9 \times \frac{3}{10} = 2\frac{7}{10}$

$3 - 2\frac{7}{10} = \frac{3}{10}$

$\frac{3}{10}$ m

Challenge

6 Use the given digits so each expression has the greatest possible value. Express the answer as a mixed number in simplest form.

(a) 4, 5, 6 $\boxed{6} \times \boxed{\dfrac{5}{4}} = 7\frac{1}{2}$

The least digit should be the denominator of the fraction. Otherwise, expressions may vary (for example (a) could be $5 \times \frac{6}{4}$).

(b) 6, 7, 8, 9 $\boxed{9} \times \boxed{\dfrac{8}{6}} + \boxed{7} = 19$

7 Use the given digits so each expression has the least possible value. Express the answer as a mixed number in simplest form.

(a) 4, 5, 6 $\boxed{4} \times \boxed{\dfrac{5}{6}} = 3\frac{1}{3}$

The greatest digit should be the denominator of the fraction.

(b) 6, 7, 8, 9 $\boxed{8} \times \boxed{\dfrac{7}{9}} + \boxed{6} = 12\frac{2}{9}$

Exercise 4 • pages 179–181

Exercise 4

Basics

1 __2__ out of 5 children are boys.

$\boxed{\dfrac{2}{5}}$ of the children are boys.

__3__ out of 5 children are girls.

$\boxed{\dfrac{3}{5}}$ of the children are girls.

2 (a) __9__ out of 21 coins are pennies.

$\boxed{\dfrac{9}{21}}$ of the coins are pennies.

__3__ out of 7 groups of coins are pennies.

$\boxed{\dfrac{3}{7}}$ of the coins are pennies.

(b) What fraction of the coins are nickels? Express the answer in simplest form.

$\frac{4}{7}$

Practice

Express all answers in simplest form.

3 What fraction of each set of stars is shaded?

(a)  $\frac{2}{7}$ (b)  $\frac{1}{5}$

(c)  $\frac{2}{3}$ (d)  $\frac{3}{4}$

(e)  $\frac{1}{2}$ (f)  $\frac{2}{3}$

(g) $\frac{3}{4}$ (h)  $\frac{5}{6}$

4

[1/6] of the shapes are squares.   [1/2] of the shapes are triangles.

[1/3] of the shapes are circles.   [2/3] of the shapes have straight sides.

5 There are 15 Mongolian gerbils and 6 fat-tailed gerbils for sale at a pet store.

(a) What fraction of the gerbils are Mongolian gerbils?

$\frac{5}{7}$

(b) What fraction of the gerbils are fat-tailed gerbils?

$\frac{2}{7}$

6 Victoria has 30 pennies, 15 nickels, and 45 dimes.

(a) What fraction of her coins are pennies?

$\frac{1}{3}$

(b) What fraction of her coins are nickels?

$\frac{1}{6}$

(c) What fraction of her coins are dimes?

$\frac{1}{2}$

Exercise 5 • pages 182–184

Exercise 5

Basics

1 There are 15 stars.

(a) $\frac{1}{5}$ of the stars are shaded.

$\frac{1}{5}$ of 15 → $\frac{15}{5}$ = [3]

$\frac{1}{5} \times 15 = \frac{1 \times \cancel{15}^3}{\cancel{5}_1}$ = [3]

(b) $\frac{4}{5}$ of the stars are shaded.

$\frac{4}{5}$ of 15 → $4 \times \frac{15}{5}$ = [12]

$\frac{4}{5} \times 15 = \frac{4 \times \cancel{15}^3}{\cancel{5}_1}$ = [12]

2 Find $\frac{5}{8}$ of 16.

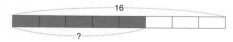

(a) 8 units → 16

1 unit → $\frac{16}{8}$ = 2

5 units → 5 × 2 = [10]

(b) $\frac{5}{8} \times 16 = \frac{5 \times \cancel{16}^2}{\cancel{8}_1}$ = [10]

Practice

Express all answers in simplest form.

3 (a) $\frac{1}{2}$ of 8 → $\frac{1}{2} \times 8$ = 4

(b) $\frac{1}{3}$ of 6 → $\frac{1}{3} \times 6$ = 2

$\frac{2}{3}$ of 6 → $\frac{2}{3} \times 6$ = 4

(c) $\frac{1}{6}$ of 12 → $\frac{1}{6} \times 12$ = 2

$\frac{5}{6}$ of 12 → $\frac{5}{6} \times 12$ = 10

(d) $\frac{1}{7} \times 21$ = 3

$\frac{3}{7} \times 21$ = 9

$\frac{5}{7} \times 21$ = 15

(e) $\frac{1}{2} \times 30$ = 15

$\frac{3}{5} \times 30$ = 18

$\frac{5}{6} \times 30$ = 25

$\frac{7}{10} \times 30$ = 21

4 Find the value of each of the following.

(a) $\frac{1}{2} \times 16$ 8

(b) $\frac{3}{4} \times 12$ 9

(c) $\frac{2}{5} \times 100$ 40

(d) $\frac{1}{9} \times 180$ 20

(e) $\frac{2}{3} \times 60$ 40

(f) $\frac{3}{8} \times 56$ 21

(g) $\frac{2}{3} \times 48$ 32

(h) $\frac{7}{10} \times 120$ 84

5 A dog has 42 teeth. $\frac{2}{7}$ of a dog's teeth are incisors, which are used for tearing meat from bones and for self-grooming. How many incisors does a dog have?

$42 \times \frac{2}{7} = 12$

12 incisors

Exercise 6 • pages 185–187

Exercise 6

Basics

1 (a) Find $\frac{1}{3}$ of 8.

$\frac{1}{3} \times 8 = \frac{8}{3}$

$= 2\frac{2}{3}$

(b) Find $\frac{2}{3}$ of 8.

$\frac{2}{3} \times 8 = \frac{2 \times 8}{3}$

$= \frac{16}{3}$

$= 5\frac{1}{3}$

2 (a) Find $\frac{1}{6}$ of 9.

$\frac{1}{6} \times 9 = \frac{9}{6}$ (with 3 and 2)

$= 1\frac{1}{2}$

(b) Find $\frac{5}{6}$ of 9.

$\frac{5}{6} \times 9 = \frac{5 \times 9}{6}$ (with 3 and 2)

$= \frac{15}{2}$

$= 7\frac{1}{2}$

3 Find $\frac{5}{9}$ of 6.

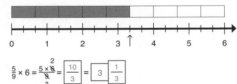

$\frac{5}{9} \times 6 = \frac{5 \times 6}{9} = \frac{10}{3} = 3\frac{1}{3}$ (with 2 and 3)

Practice

Express answers 1 or greater as whole or mixed numbers. Use simplest form.

4 Find the value of each of the following.

(a)

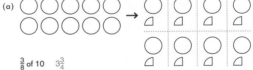

$\frac{3}{8}$ of 10 $3\frac{3}{4}$

(b)

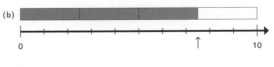

$\frac{3}{4}$ of 10 $7\frac{1}{2}$

5. Find the value of each of the following.

(a) $\frac{1}{2}$ of 9 $4\frac{1}{2}$

(b) $\frac{1}{3}$ of 10 $3\frac{1}{3}$

(c) $\frac{2}{9} \times 4$ $\frac{8}{9}$

(d) $\frac{5}{9} \times 3$ $1\frac{2}{3}$

(e) $\frac{3}{4} \times 30$ $22\frac{1}{2}$

(f) $\frac{7}{8} \times 20$ $17\frac{1}{2}$

(g) $\frac{3}{7} \times 18$ $7\frac{5}{7}$

(h) $\frac{7}{10} \times 15$ $10\frac{1}{2}$

6. Does $\frac{5}{8}$ of a 12 lb bag of beans weigh more or less than twelve $\frac{5}{8}$ lb bags of beans?

They weigh the same.

Exercise 7 • pages 188–191

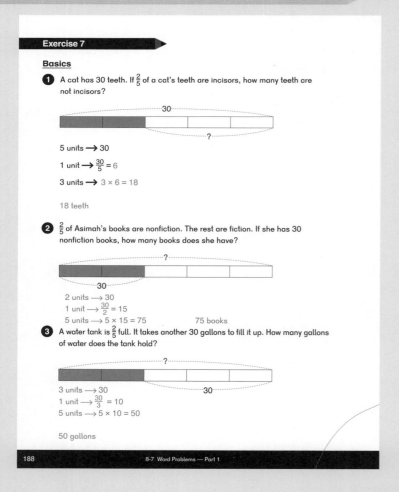

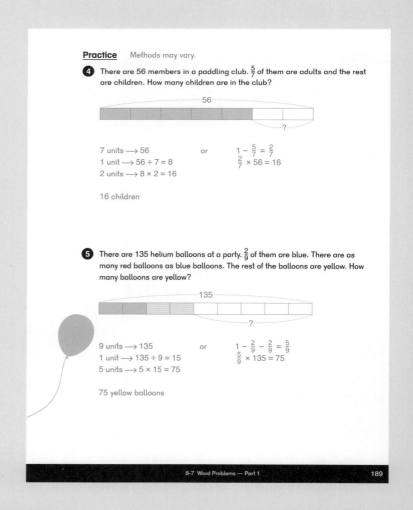

6. Gary traveled by train for 400 miles of his trip. He took a bus for the remaining $\frac{3}{8}$ of the trip. How many miles was his trip?

5 units ⟶ 400
1 unit ⟶ 400 ÷ 5 = 80
8 units ⟶ 80 × 8 = 640
640 miles

7. Lee made a brick mailbox using two colors of bricks. $\frac{5}{12}$ of the bricks were light red and the rest were dark red. If he used 85 light red bricks, how many bricks did he use in all?

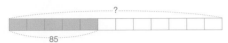

5 units ⟶ 85
1 unit ⟶ 85 ÷ 5 = 17
12 units ⟶ 12 × 17 = 204
204 bricks

Challenge

8. After $\frac{3}{5}$ of a bag of flour was used to make bread, there were 9 kg of flour left. How many kilograms of flour were in the bag at first?

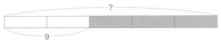

2 units ⟶ 9
1 unit ⟶ 9 ÷ 2 = $4\frac{1}{2}$
5 units ⟶ 5 × $4\frac{1}{2}$ = $22\frac{1}{2}$
$22\frac{1}{2}$ kg

9. If $\frac{7}{15}$ of a number is 35, what is $\frac{3}{5}$ of the number?

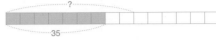

$\frac{3}{5} = \frac{9}{15}$
7 units ⟶ 35
1 unit ⟶ 35 ÷ 7 = 5
9 units ⟶ 9 × 5 = 45
45

Exercise 8 • pages 192–195

Exercise 8

Basics

1. Anna spent 21 days on a vacation. 14 of them were spent at the beach and the remaining days were spent hiking in the mountains. What fraction of her vacation was spent in the mountains?

21 − 14 = 7

7 out of 21 days were spent in the mountains.

$\frac{7}{21} = \frac{1}{3}$

$\frac{1}{3}$ of her vacation was spent in the mountains.

2. On the way home from the park, Jody ran for $\frac{3}{5}$ of the way, jogged for $\frac{1}{10}$ of the way, and walked the remaining 360 m. How far was the park from her home in meters?

$\frac{3}{5} = \frac{6}{10}$

3 units ⟶ 360
1 unit ⟶ 360 ÷ 3 = 120
10 units ⟶ 10 × 120 = 1,200
1,200 m

3. Fadiya had 92 tropical fish. She gave 8 of them to her friend, and then sold $\frac{2}{7}$ of the remaining fish. How many fish does she have now?

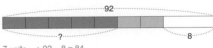

7 units ⟶ 92 − 8 = 84
1 unit ⟶ 84 ÷ 7 = 12
5 units ⟶ 5 × 12 = 60
60 fish

Practice

4. Jack is driving from town to his farm 36 miles away. 25 miles of the road are paved, 7 miles are gravel, and the remaining miles are dirt. What fraction of the trip is on a dirt road?

36 − 25 − 7 = 4

$\frac{4}{36} = \frac{1}{9}$

$\frac{1}{9}$ of the trip

Teacher's Guide 4A Chapter 8

5. Aurora baked some cookies. She gave 12 cookies to her family, and then sold $\frac{7}{9}$ of the cookies at a bake sale. She had 8 cookies left over. How many cookies did she bake?

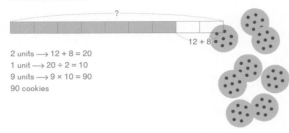

2 units ⟶ 12 + 8 = 20
1 unit ⟶ 20 ÷ 2 = 10
9 units ⟶ 9 × 10 = 90
90 cookies

6. On a math team, 6 students are from the third grade, 9 students are from the fourth grade, and $\frac{4}{9}$ of the students are from the fifth grade. How many students are from the fifth grade?

5 units ⟶ 6 + 9 = 15
1 unit ⟶ 15 ÷ 5 = 3
4 units ⟶ 4 × 3 = 12
12 students

7. Land area is measured in acres. On a farm, $\frac{1}{12}$ of the land is used to grow clover, $\frac{1}{4}$ is used to grow wheat, $\frac{1}{3}$ is used to grow rye, and the remaining 8 acres is used to grow oats. How many acres is the land that is being planted?

4 units ⟶ 8
1 unit ⟶ 8 ÷ 4 = 2
12 units ⟶ 12 × 2 = 24
24 acres

Challenge

8. There are twelve flags spaced equally along a track. Runners start at the first flag. A runner reaches the 8th flag in 8 seconds. If he runs the whole time at the same speed, in how many seconds will he reach the 11th flag?

There are 7 intervals from the 1st to the 8th flag, and 10 from the 1st flag to the 11th flag. So he ran each segment in $\frac{8}{7}$ seconds, and 10 segments in $10 \times \frac{8}{7} = \frac{80}{7} = 11\frac{3}{7}$ seconds.

$11\frac{3}{7}$ seconds

Exercise 9 • pages 196–200

Exercise 9

Check

Express answers 1 or greater as whole or mixed numbers. Use simplest form.

1. Multiply.

   (a) $15 \times \frac{1}{5}$ 3
   (b) $12 \times \frac{2}{3}$ 8
   (c) $6 \times \frac{5}{9}$ $3\frac{1}{3}$
   (d) $4 \times \frac{7}{30}$ $\frac{14}{15}$
   (e) $\frac{3}{5} \times 10$ 6
   (f) $\frac{7}{8} \times 40$ 35
   (g) $\frac{3}{7} \times 8$ $3\frac{3}{7}$
   (h) $\frac{13}{20} \times 480$ 312

2. (a) $24 \times \boxed{\frac{1}{3}} = 8$ (b) $36 \times \boxed{\frac{1}{4}} = 9$

3. A nail is $\frac{5}{8}$ inches long. How many inches are 7 nails placed end-to-end?

   $7 \times \frac{5}{8} = 4\frac{3}{8}$

   $4\frac{3}{8}$ inches

4. Because of the lower gravity of the moon, a person weighs about $\frac{4}{25}$ as much on the moon as on Earth. About how much would a 75 pound person weigh on the moon?
   $75 \times \frac{4}{25} = 12$
   12 pounds

5. Because of the lower gravity of Mars, a person who weighs 200 pounds on Earth would weigh 76 pounds on Mars.

   (a) A weight on Mars is what fraction of a weight on Earth?
   $\frac{76}{200} = \frac{19}{50}$

   $\frac{19}{50}$

   (b) How much would a 75 pound person weigh on Mars?

   $75 \times \frac{19}{50} = 28\frac{1}{2}$

   $28\frac{1}{2}$ pounds

6. There are some balloons at a party. $\frac{2}{9}$ of the balloons are black, $\frac{1}{3}$ of them are orange, and the remaining 16 balloons are yellow. How many balloons are there?

$\frac{1}{3} = \frac{3}{9}$

4 units ⟶ 16
1 unit ⟶ 16 ÷ 4 = 4
9 units ⟶ 4 × 9 = 36
36 balloons

7. On a farm, there are 60 sheep. 16 are Leicester sheep. There are twice as many Merino sheep as Leicester sheep. The rest are Dorset sheep. What fraction of the sheep are Dorset sheep?

Merino: 2 × 16 = 32
Dorset: 60 − 32 − 16 = 12
$\frac{12}{60} = \frac{1}{5}$

$\frac{1}{5}$ of the sheep

8. Amelia has 24 toy cars. She gave $\frac{3}{4}$ of her cars to Carter. Carter gave $\frac{2}{3}$ of the cars he got from Amelia to Eli. Eli gave $\frac{1}{2}$ of the cars he got from Carter to Grace. How many cars did Grace get from Eli?

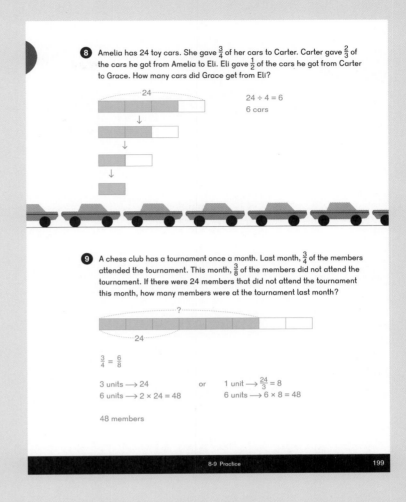

24 ÷ 4 = 6
6 cars

9. A chess club has a tournament once a month. Last month, $\frac{3}{4}$ of the members attended the tournament. This month, $\frac{3}{8}$ of the members did not attend the tournament. If there were 24 members that did not attend the tournament this month, how many members were at the tournament last month?

$\frac{3}{4} = \frac{6}{8}$

3 units ⟶ 24   or   1 unit ⟶ $\frac{24}{3}$ = 8
6 units ⟶ 2 × 24 = 48   6 units ⟶ 6 × 8 = 48

48 members

Challenge

10. 8 stakes are spaced evenly apart in a garden. If the distance between two of them is $\frac{4}{7}$ m, how many meters is the distance from the first stake to the last stake?

There are 7 intervals.
7 × $\frac{4}{7}$ = 4
4 meters

11. $\frac{1}{5}$ of $\frac{1}{3}$ of $\frac{1}{2}$ of a number is 25. What is the number?

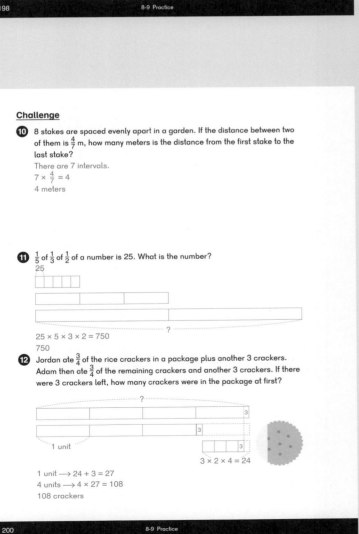

25 × 5 × 3 × 2 = 750
750

12. Jordan ate $\frac{3}{4}$ of the rice crackers in a package plus another 3 crackers. Adam then ate $\frac{3}{4}$ of the remaining crackers and another 3 crackers. If there were 3 crackers left, how many crackers were in the package at first?

3 × 2 × 4 = 24

1 unit ⟶ 24 + 3 = 27
4 units ⟶ 4 × 27 = 108
108 crackers

Notes

Chapter 9 Line Graphs and Line Plots — Overview

Suggested number of class periods: 5–6

	Lesson	Page	Resources	Objectives
	Chapter Opener	p. 281	TB: p. 211	Investigate and recall graphs.
1	Line Graphs	p. 282	TB: p. 212 WB: p. 201	Interpret change over time in bar and line graphs.
2	Drawing Line Graphs	p. 286	TB: p. 218 WB: p. 205	Draw and display data in a line graph.
3	Line Plots	p. 288	TB: p. 221 WB: p. 208	Interpret and create line plots.
4	Practice	p. 291	TB: p. 226 WB: p. 211	Practice reading and creating line graphs and line plots.
	Review 2	p. 293	TB: p. 228 WB: p. 215	Review content from Chapters 1–9.
	Workbook Solutions	p. 296		

Chapter 9 Line Graphs and Line Plots

In Dimensions Math 3A Chapter 7, students learned to:

- Interpret and create scaled picture graphs and bar graphs.
- Create bar graphs using data from tables.

In this chapter, students will learn to read and interpret line graphs and line plots to see trends in data. They will also create line graphs from a set of data.

Data

Facts and statistics that are gathered for analysis are called data.

Tables

A table is an organized way to present data in rows and columns. Students will compare the information in tables to the same information shown in graphs. They will create a graph from a table.

Line Graphs

A line graph is generally used to show change over time. The horizontal axis usually shows each time period. The vertical axis shows the data that is collected for that time period. Data points are plotted on a grid and connected by line segments. Line graphs provide an easy way to track characteristics and trends of data over a period of time.

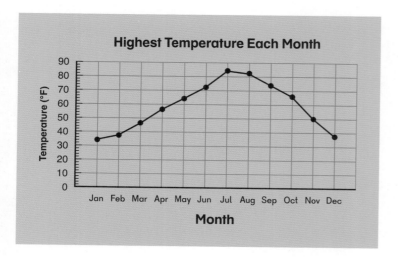

Line Plots

Line plots, or dot plots, are used to show the frequency of data values along a number line. They are generally used for small data sets and are useful for seeing how data is distributed. Each point or dot on the graph represents a data value.

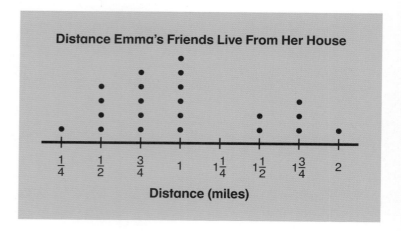

The scale on a line plot typically begins with the least value of the data and ends with the greatest.

Encourage students to compare the different ways of presenting data in bar graphs, tables, line graphs, and line plots.

There are minimal supplemental activities in this chapter, as students should spend most of each lesson creating and interpreting the graphs and line plots. Extend the lessons by asking students to think critically about what type of data they want to represent and what type of graph works best to present that data.

After students complete the chapter in the textbook, have them continue to practice creating and reading graphs with content from other school subjects. For example, graph data from science experiments.

Chapter 9 Line Graphs and Line Plots

Materials

- Dry erase markers
- Newspapers or magazines
- Ruler or straightedge
- Set of data
- Whiteboards

Blackline Masters

- Graph paper
- Line Graph

Activities

Minimal games and activities are included in this chapter as students will be creating graphs. Use activities from prior chapters as needed, for additional practice with fraction skills.

Notes

Chapter Opener

Objective

- Investigate and recall graphs.

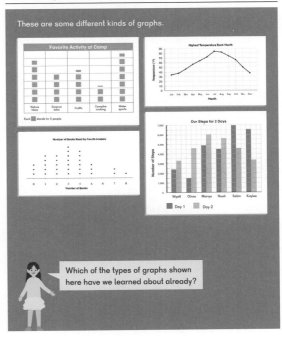

Use the **Chapter Opener** as assessment of prior knowledge of graphs.

Discuss the textbook page and have students recall graphs they have seen before.

Continue to Lesson 1 or extend the **Chapter Opener** to a full lesson with an **Activity**.

Activities

▲ Newspaper Graphs

Materials: Newspapers or magazines

Provide (or ask students to bring) newspapers or magazines. Have students search their materials for graphs and have them share what information is shown in the graphs.

▲ Make a Graph

Materials: Set of data, Graph Paper (BLM)

Provide students with a set of data that can be graphed and ask them to create a way of graphing the data. They can use any type of graph that is already familiar to them.

Suggested data to use:

- Number of students eating hot lunch each day for one week.
- Number of books checked out of the school library each day for one week.

Lesson 1 Line Graphs

Objective

- Interpret change over time in bar and line graphs.

Think

Discuss the information recorded in Dion's table in **Think** as well as his comments about statistics and data.

Ask students about the parts of the table:

- "What is the title of the graph?" (Highest Temperature Each Month)
- "What are the legends on the vertical and horizontal lines?" (Temperature and Month)
- "How many values are given?" (12, one for each month)

Have students compare the information in the table to the bar graph.

Possible questions to ask students:

- "How is the data from the table represented?"
- "Why is it helpful to display the data in the table in a graph?"
- "What are the labels on the graph?"
- "How are the intervals labeled on the vertical number line?"
- "Which month had the lowest high temperature recorded? Is it easy to see from the graph?"
- "Which month had the greatest high temperature recorded?"

Finally, discuss the **Think** question, "What do you notice about how the high temperature changes from month to month and during the whole year?"

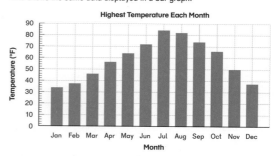

Learn

Introduce line graphs to students. Students can see that since the height of a bar is the same as a data value, a dot placed at the top of each bar also represents a data value. The dot is at the intersection of a vertical line (the month), and a horizontal line (the temperature).

Students should see that the horizontal axis shows the months in a year, and that the line connecting data points shows the changes in temperature over time.

Discuss the differences between the bar graph and the line graph. Ask students why one graph might be more helpful for certain data.

One type of graph might be more helpful than the other depending on the data that is represented and the analysis it is being used for.

Point out to students the range of the data. The highest point represents the greatest data value.

Ask students why the temperature is marked only in multiples of 10 along the vertical axis.

Ask students, "Looking at the graph, how can we tell between which months the highest temperature increases and decreases?" (When the line goes up, the highest temperatures are increasing, when the line goes down, the highest temperatures are decreasing.)

Dion points out that the trends, or changes, are easy to see on the graph.

Ask students:

- "Looking only at the graph, can we find the exact highest temperature for each month?"
- "Is it easier to find the exact highest temperature from the table or the graph?"

Mei points out that it is easier to use a table to find exact values.

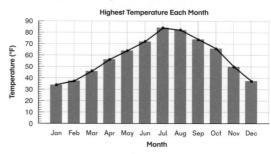

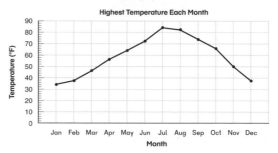

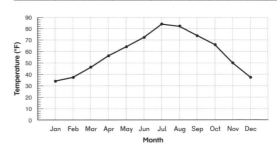

Do

1 (a), (b) The horizontal axis represents the months. The vertical axis represents the data for the given time period.

(c) Students can count that there are five total marks from 0 to 10. 10 ÷ 5 = 2, so each shorter black mark denotes 2°F in temperature.

(e), (g), (i) Students should see that this information is easier to find on the table. For example, it is easier to see the difference between 34°F and 38°F on the table than by counting tick marks.

(f), (h) It is easier to use the graph to see that the sharpest rise in the high temperature occurred between June and July. The sharpest rise looks steeper on the graph. Similarly, it is also easier to use the graph to see the sharpest drop in temperature.

2 No table is provided. Dashed lines are included on the graph to make it easier for students to read.

3 From the table, students should see that there is an increase in temperature from 7:00 am to 4:00 pm, followed by a decrease.

Students should match this trend to the corresponding graph.

There are no axes on the graph, so we do not actually know the scale and cannot determine the exact values from the graph.

Since the vertical axis is a number line, differences in height between two data points correspond proportionately to differences between those data values. A steeper line corresponds to a greater difference in the values.

Note: Students will create a line graph in the next lesson.

Exercise 1 • page 201

Lesson 2 Drawing Line Graphs

Objective
- Draw and display data in a line graph.

Lesson Materials
- Line Graph (BLM)
- Ruler or straightedge

Think

Have students discuss the information given in the table.

Ask students:

- "How can we draw a line graph?"
- "What data goes on the vertical and horizontal axes?"
- "What would be the most efficient scale to use?"

Learn

Follow the directions in **Learn** to represent the given data on the Line Graph (BLM).

Ensure students use a ruler or straightedge to connect the data points.

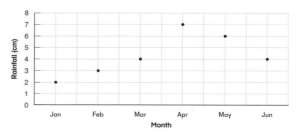

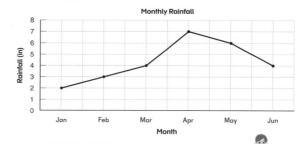

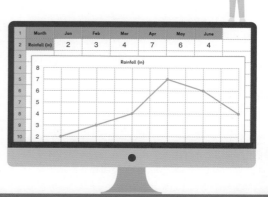

Do

① – ② Prompt students to think about Sofia's question before drawing line graphs for each table.

Choose a scale so that all the data points are visible on the graph and differences between the points are easy to see.

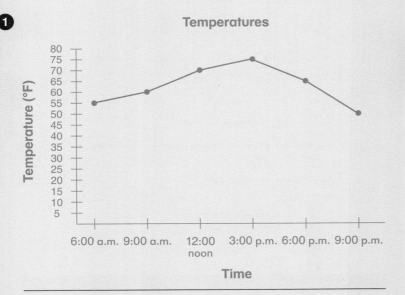

Do

① The table below shows the temperature in degrees Fahrenheit in a certain town from 6:00 a.m. to 9:00 p.m.

Time	6:00 a.m.	9:00 a.m.	12:00 noon	3:00 p.m.	6:00 p.m.	9:00 p.m.
Temp (°F)	55	60	70	75	65	50

Draw a line graph on graph paper to display the data.

 What scale should we use on the vertical axis?

② The table below shows the number of shoppers at a store during each hour.

Time	\multicolumn{5}{c}{a.m.}	\multicolumn{7}{c}{p.m.}
Time	7	8	9	10	11	12	1	2	3	4	5	6	7
Shoppers	19	36	28	52	69	86	73	52	47	55	68	34	23

(a) Draw a line graph on graph paper to display the data.

(b) When does the store need the most cashiers?
Answer will vary. Around noon and 5 p.m.

(c) When would be a good time to shop to avoid a crowded store?
Answers will vary. Early morning and late evening.

Exercise 2 • page 205

9-2 Drawing Line Graphs

Exercise 2 • page 205

Activity

▲ If technology is available, have students create a line graph using a spreadsheet program and data from **① – ②**.

© 2019 Singapore Math Inc. Teacher's Guide 4A Chapter 9

Lesson 3 Line Plots

Objective
- Interpret and create line plots.

Lesson Materials
- Graph paper (BLM)

Think

Discuss the table and the three questions with students.

Alex asks, "How can we organize this data to easily answer question (c)?"

Learn

Tell students that a line plot makes it easier to see the range of data and the number of occurrences of data. In the line plot for each of the children's ages, it is easy to see that the most common age is 9.

Discuss the horizontal axis with students. They should notice that the dots are evenly spaced vertically as if there were a vertical axis.

Ask students:

- "Why is there no vertical axis on a line plot?"
- "Why can we start with the number 6 on the horizontal axis?"
- "Why can we end with the number 12 on the horizontal axis?"

The horizontal axis is a number line representing data on ages. We do not need to begin at 0. It is best to begin and end with the minimum and maximum values of the ages.

Mei points out that it is easy to see that most children at the party were 9 or 10 years old, and no children were 8 years old. Ask students why this might be so.

Lesson 3
Line Plots

Think

The table below shows the names and ages of children at Mei's birthday party.

Name	Age (years)
Dion	9
Pablo	6
Kawai	10
Emma	9
Alex	10
Kaitlyn	7
Sofia	9
Mei	10
Ximena	12
Tomas	9

(a) How many children are at the party?

(b) What is the difference between the age of the youngest child and the age of the oldest child?

(c) What ages are the most common?

How can we organize the data to make it easier to read?

Learn

We can use a **line plot** to organize the data.

Ages of Children at Mei's Birthday Party

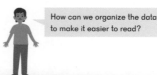

(a) There were 10 children at the party.

(b) The youngest child is 6 years old.

The oldest child is 12 years old.

The difference between the age of the youngest child and the age of the oldest child is 6 years.

(c) The most common ages are 9 and 10 .

A line plot makes it easier to see clusters and gaps in the values for small sets of data. They are also sometimes called dot plots.

Do

1 (b) Students can count, or subtract from the answer found in (a).

2 The number line for this line plot includes fractions. Ask students why fractions can be included in this data set and not in the data set in **Learn**.

Ask students what other information they can take away from the graph. Example answers:

- 3 out of 32 students read no books last month.
- 3 out of 32 students read 7 or 8 books last month.

③ Provide students with Graph Paper (BLM). Students should think about how to draw the number line. They need to determine the least and greatest data value and be able to fit that range on their pages, and also be able to show fourths.

Since the least amount of rainfall is 0 inches, the number line should start with 0.

The data is listed in fractional amounts of fourths and halves. Ask students what scale would work best on the line plot.

Students should see that $\frac{1}{4}$ increments up to $2\frac{1}{2}$ ($2\frac{2}{4}$) inches would cover each data point:

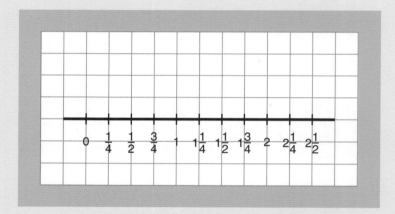

When students have the number line completed, they can begin plotting the data.

③

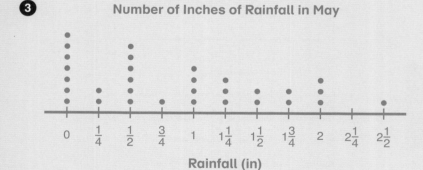

Exercise 3 • page 208

---

③ Mei collected data on the amount of rainfall each day in May. She measured up to each $\frac{1}{4}$ inch interval.

$\frac{1}{4}$ means it rained up to $\frac{1}{4}$ inch. $\frac{1}{2}$ means it rained more than $\frac{1}{4}$ inch up to but less than $\frac{1}{2}$ inch, and so on.

Number of Inches of Rainfall in May					
		2	2	$1\frac{3}{4}$	$\frac{1}{2}$	$\frac{1}{4}$
$\frac{1}{2}$	$\frac{1}{2}$	$\frac{1}{4}$	$\frac{3}{4}$	0	0	0
0	1	1	2	$1\frac{1}{2}$	$\frac{1}{2}$	$\frac{1}{2}$
$\frac{1}{2}$	1	$2\frac{1}{2}$	$1\frac{1}{4}$	1	$1\frac{1}{2}$	0
$1\frac{1}{4}$	$1\frac{1}{4}$	$1\frac{3}{4}$	0	0		

(a) Draw a line plot on graph paper to show the information. First draw a horizontal axis. Find the least and greatest values and decide on a scale along the axis. Then draw dots for each data value. Write a title for your graph.

(b) On how many days did it rain?
24 days

(c) On how many days did it rain 1 or more inches?
15 days

(d) On how many days did it rain less than 1 inch?
9 days

(e) What was the most common amount of rain in a day when it did rain?
$\frac{1}{2}$ in

Exercise 3 • page 208

Lesson 4 Practice

Objective

- Practice reading and creating line graphs and line plots.

Lesson Materials

- Graph paper (BLM)

Remind students to think about the scale they should use before drawing the graph or line plot.

After students complete the **Practice** in the textbook, have them continue to practice interpreting data from graphs with content from other school subjects.

1

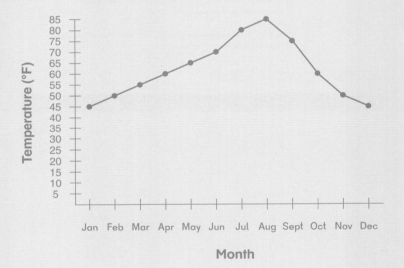

Lesson 4 Practice — P 4

1

The table shows the monthly high temperatures from January to December in a given year in Seattle, Washington, rounded to the nearest 5 degrees Fahrenheit.

Month	Jan	Feb	Mar	Apr	May	Jun	Jul	Aug	Sep	Oct	Nov	Dec
Temp (°F)	45	50	55	60	65	70	80	85	75	60	50	45

Create a line graph to show the data.

(a) Which month had the hottest day?
August
(b) Which months had the lowest high temperature?
January and December
(c) What is the difference in the highest temperature and the lowest temperature?
85 − 45 = 40; 40 °F
(d) How many months had a day with a high temperature above 60 degrees?
5 months
(e) How many months had no days warmer than 60 degrees?
7 months

2 Foot Length of Some Fourth Graders

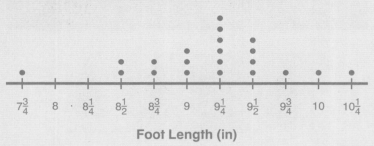

Foot Length (in)

Activity

Students can revisit the newspaper graphs from the **Chapter Opener** and identify the type of each graph, its labels, axes, and data.

▲ **Newspaper Graphs**

Materials: Newspapers or magazines

Provide (or have students bring) newspapers and magazines. Have students search their materials for graphs and have them share what information is shown in the graph.

Exercise 4 • page 211

2

Some fourth grade students measured the length of their feet to the nearest $\frac{1}{4}$ inch.

Foot Length (inches)							
$9\frac{1}{2}$	$9\frac{1}{4}$	$8\frac{3}{4}$	$9\frac{3}{4}$	9	$9\frac{1}{2}$	$8\frac{3}{4}$	$9\frac{1}{4}$
$9\frac{1}{4}$	$9\frac{1}{2}$	9	$7\frac{3}{4}$	$8\frac{1}{2}$	$10\frac{1}{4}$	$9\frac{1}{4}$	9
10	$8\frac{1}{2}$	$9\frac{1}{4}$	$9\frac{1}{4}$	$9\frac{1}{2}$			

Draw a line plot to show the data.

(a) How many measurements were recorded?
    21
(b) What is the length of the longest foot?
    $10\frac{1}{4}$ in
(c) What is the length of the shortest foot?
    $7\frac{3}{4}$ in
(d) What is the difference between the length of the longest foot and the length of the shortest foot?
    $10\frac{1}{4} - 7\frac{3}{4} = 2\frac{1}{2}$; $2\frac{1}{2}$ in
(e) Sock sizes 3–5 fit feet that are from $8\frac{1}{2}$ to $9\frac{1}{8}$ inches long. How many of these students would wear these sizes?
    7 students
(f) Sock sizes 5–7 fit feet that are from $9\frac{1}{8}$ to $9\frac{13}{16}$ inches long. How many of these students would wear these sizes?
    11 students

Exercise 4 • page 211

Review 2

Objective

- Review content from Chapters 1–9.

Reviews are important for reinforcing skills. They also help teachers identify where remediation of specific topics might be necessary.

These pages can be used to review content before doing a mid-year cumulative assessment. This review can also serve as an assessment.

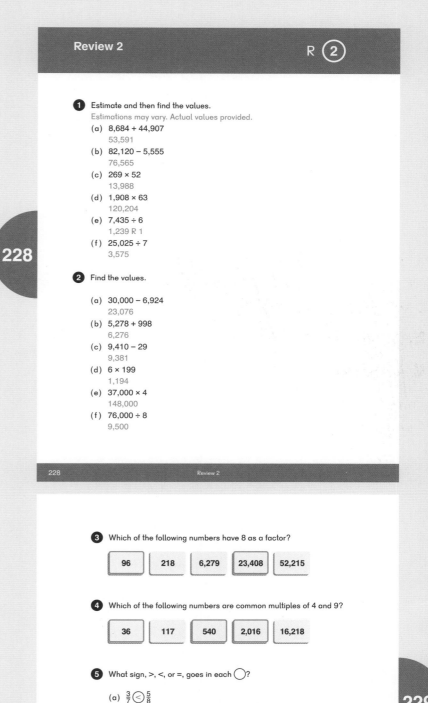

6 Express each of the following as a proper fraction or mixed number. Express each answer in its simplest form.

(a) $8 \div 10$ $\frac{4}{5}$ (b) $12 \div 18$ $\frac{2}{3}$

(c) $38 \div 7$ $5\frac{3}{7}$ (d) $21 \div 6$ $3\frac{1}{2}$

(e) $36 \div 8$ $4\frac{1}{2}$ (f) $100 \div 30$ $3\frac{1}{3}$

7 Find the value. Express each answer in its simplest form.

(a) $\frac{6}{7} - \frac{2}{7}$ $\frac{4}{7}$

(b) $4 - \frac{3}{8}$ $3\frac{5}{8}$

(c) $\frac{5}{6} + \frac{5}{6}$ $1\frac{2}{3}$

(d) $\frac{7}{12} + \frac{3}{4}$ $1\frac{1}{3}$

(e) $3\frac{5}{9} + \frac{7}{9}$ $4\frac{1}{3}$

(f) $6\frac{5}{8} - \frac{1}{4}$ $6\frac{3}{8}$

(g) $8\frac{5}{6} + 4\frac{5}{12}$ $13\frac{1}{4}$

(h) $7\frac{1}{6} - 3\frac{2}{3}$ $3\frac{1}{2}$

(i) $2\frac{3}{5} + \frac{1}{10} + 4\frac{1}{5}$ $6\frac{9}{10}$

230

8 Find the product. Express each answer in its simplest form.

(a) $2 \times \frac{1}{5}$ $\frac{2}{5}$

(b) $6 \times \frac{1}{8}$ $\frac{3}{4}$

(c) $\frac{2}{3} \times 3$ 2

(d) $\frac{3}{4} \times 12$ 9

(e) $\frac{4}{5} \times 20$ 16

(f) $15 \times \frac{3}{9}$ 5

(g) $\frac{2}{9} \times 6$ $1\frac{1}{3}$

(h) $\frac{2}{3} \times 20$ $13\frac{1}{3}$

(i) $\frac{5}{3} \times 10$ $16\frac{2}{3}$

9 There are 25 waterfowl in a pond. 10 of them are ducks and the rest are geese. What fraction of the waterfowl are geese? (Give your answer in simplest form.)
Geese: $25 - 10 = 15$; $\frac{15}{25} = \frac{3}{5}$
$\frac{3}{5}$ of the waterfowl

10 Shoshana had $\frac{8}{9}$ m of ribbon. She cut off two $\frac{1}{3}$ m long pieces to wrap two presents. How long is the piece she has left?
Pieces used: $\frac{1}{3} + \frac{1}{3} = \frac{2}{3}$; Piece left: $\frac{8}{9} - \frac{2}{3} = \frac{2}{9}$; $\frac{2}{9}$ m

231

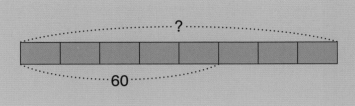
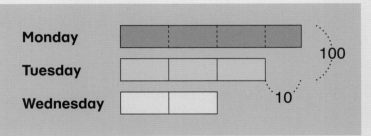

Exercise 5 • page 215

Brain Works

★ Number Patterns

(a) 2 + 4 + 6 + 4 + 2 =
(b) 1 + 3 + 5 + 7 + 5 + 3 + 1 =
(c) 1 + 2 + 3 + ... + 9 + 10 =
(d) 2 + 4 + 6 + ... + 16 + 18 + 20 =
(e) 1 + 3 + 5 + 7 + ... + 17 + 19 + 21 =

Students should see that they can group the numbers to make the calculation easier. Here are some examples.

(a) (2 + 4) + 6 + (4 + 2) = 3 × 6 = 18
(b) 1 + 7 + 3 + 5 + 5 + 3 + 1 = 3 × 8 + 1 = 25
(c) 1 + 10, 2 + 9, 3 + 8, 4 + 7, 5 + 6 make 5 groups of 11, or 55
(d) 2 + 20, 4 + 18, 6 + 16, 8 + 14, 10 + 12 make 5 × 22 = 110
(e) 1 + 21, 3 + 19, 5 + 17, 7 + 15, 9 + 13 make 5 × 22 + 11 = 121

© 2019 Singapore Math Inc.   Teacher's Guide 4A Chapter 9   295

Exercise 1 • pages 201–204

Chapter 9 Line Graphs and Line Plots

Exercise 1

Basics

1. This table shows the number of days it snowed each month from October to June at Cloud Ski Resort. A graph of the same data is partially completed.

Month	Oct	Nov	Dec	Jan	Feb	Mar	Apr	May	Jun
Number of Days of Snow	0	7	8	26	16	20	7	3	1

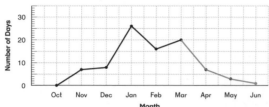

(a) Complete the graph.

(b) Which three months had the most days of snowfall?
January, February, March

(c) The greatest change in the number of days of snowfall was between which two months? Was this change an increase or a decrease?
December to January; Increase

Practice

2. This table shows the greatest snow depth, in inches, each month from October to June at a ski resort. A graph of the same data is partially completed.

Month	Oct	Nov	Dec	Jan	Feb	Mar	Apr	May	Jun
Depth (in)	0	20	63	155	180	168	160	155	80

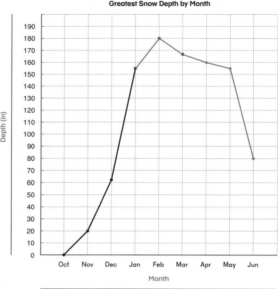

(a) Plot the rest of the points and connect them with lines.

(b) What does the vertical axis represent? Label it.
Depth of snow in inches

(c) What does the horizontal axis represent? Label it.
Month

(d) How many inches does each increment between tick marks represent?
10 inches

(e) In which month was the snow the deepest?
February

(f) Between which two months was the sharpest increase in snow depth?
December to January

(g) By how many inches did the snow depth fall between January and June?
75 inches

(h) Which months might be the best time to plan a ski trip when there is fresh snow?
December or January

3. This table shows the number of people that visited a museum each day for one week.

Day	Mon	Tues	Wed	Thur	Fri	Sat	Sun
Number of People	35	25	12	20	42	60	55

Which three of the following graphs could be accurate representations of the data?   B, C, and F. Scales differ on each of them.

A

B

C

D

E

F

Exercise 2 • pages 205–207

Exercise 2

Basics

1 The following table shows the number of children each week at the 8 sessions of a karate class.

Session	1	2	3	4	5	6	7	8
Number of Children	30	25	20	23	17	15	12	20

Complete the graph below to show this information.   Title may vary.

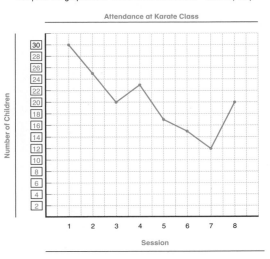

Practice

2 Misha recorded the weight of her Shih Tzu puppy, Sofie, every 2 weeks for the first 36 weeks. The data is shown in the table below.

(a) Complete the line graph on the next page. Include a title, label the axes, and label the increments.

(b) At about how many weeks did Sofie's growth rate start slowing down?
24 weeks

(c) At about how many weeks old was Sofie full grown?
30 weeks

(d) From the graph, estimate Sofie's weight at 17 weeks.
72 ounces

(e) What would be an expected weight for Sofie at 1 year?
around 98 ounces

Weeks	Weight (Ounces)
Birth	6
2	16
4	24
6	29
8	39
10	45
12	53
14	60
16	70
18	74
20	81
22	86
24	93
26	95
28	96
30	98
32	99
34	98
36	98

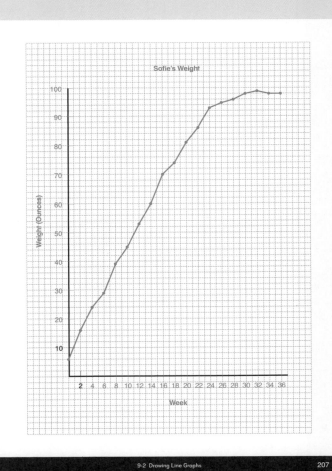

Exercise 3 • pages 208–210

Exercise 3

Basics

1. A visitor at Yellowstone National Park recorded the duration of the eruptions of the Old Faithful Geyser to the nearest fourth of a minute.

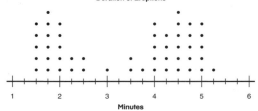

Duration of Eruptions

(a) How many eruptions did he record?
50

(b) The eruptions lasted from $1\frac{1}{2}$ minutes to $5\frac{1}{4}$ minutes.

(c) Which two time durations were most of the eruptions?
$1\frac{3}{4}$ minutes and $4\frac{1}{2}$ minutes

(d) How many time durations were less than $3\frac{1}{4}$ minutes long?
21

Practice

2. Jaiden asked some students how long it took them to get to school, rounded to the nearest 5 minutes, and listed their answers.

10	20	5	25	15	10	30	15	25	20	25	40	30	15	35
20	25	35	15	30	20	40	10	30	15	10	50	45	20	25

(a) Use this data to complete the line plot below.

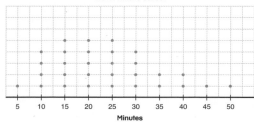

Travel Time to School

(b) What are the most common travel times?
15, 20, and 25 minutes

(c) What is the difference between the longest and shortest travel times?
45 minutes

(d) How many students took less than half an hour to travel to school?
20

(e) What fraction of the students took more than half an hour to travel to school?
$\frac{1}{5}$

3. The students in a fourth grade class were asked to write down the number of hours they slept to the nearest fourth of an hour.

9	$8\frac{3}{4}$	6	$9\frac{1}{4}$	$8\frac{3}{4}$	8	$7\frac{1}{4}$	$9\frac{1}{4}$
$9\frac{1}{2}$	10	$9\frac{1}{2}$	$6\frac{1}{2}$	$9\frac{1}{2}$	9	$9\frac{1}{4}$	$10\frac{1}{4}$
9	$7\frac{1}{4}$	8	$8\frac{1}{2}$	9	$7\frac{3}{4}$	9	11
10	$9\frac{1}{4}$	$8\frac{1}{2}$	$8\frac{3}{4}$	$10\frac{1}{4}$	$9\frac{3}{4}$	$9\frac{1}{2}$	$10\frac{3}{4}$

(a) Complete the line plot below.

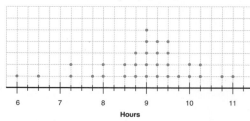

Hours of Sleep for Fourth Graders

(b) How many students are included in the data set?
32 students

(c) It is recommended that children between 6 and 13 years old get at least 9 hours of sleep. How many students did not get at least 9 hours of sleep?
12 students

(d) What fraction of the students got at least 9 hours of sleep?
$\frac{5}{8}$

Exercise 4 • pages 211–214

Exercise 4

Check

① A teacher recorded the number of words some students could type before and after a keyboarding class. These two line plots show the two data sets.

Before the Course

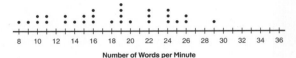

Number of Words per Minute

After the Course

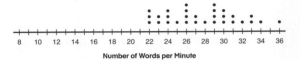

Number of Words per Minute

(a) How many students were in the class?
  30 students

(b) What was the difference between the least and the greatest number of words students could type in one minute at the beginning of the course?
  21 words per minute

(c) What was the difference between the least and the greatest number of words students could type in one minute at the end of the course?
  14 words per minute

(d) Based on the data, was the class helpful? Explain.
  The class was helpful. Answers will vary. It is not possible to determine from this data whether all students improved, but most of them did.

② Jacob's and Sara's heights to the nearest inch was recorded on their birthdays every year from ages 2 to 20. The data is shown in the table below. On the next page, Jacob's data has been graphed. The vertical axis below 30 has been shortened since there are no data values for height less than 30.

(a) Write a title and label the axes on the graph.

(b) Use a different color to graph the data for Sara. Indicate which line is for which person.

(c) From the graph, estimate Jacob's height when he was $3\frac{1}{2}$ years old.
  38 inches

(d) From the graph, estimate Sara's height when she was $11\frac{1}{2}$ years old.
  60 inches

(e) At about what age did Sara stop growing as quickly?
  About 12 years

(f) At about what age did Jacob stop growing as quickly?
  About 17 years

(g) What other observations can you make from the data?
Answers will vary. Examples: Sara was shorter to start with, grew more quickly than Jacob and so was taller at the same age for a while, but stopped growing sooner and by age 20 was shorter than Jacob.

Age (years)	Height (in) Jacob	Height (in) Sara
2	35	31
3	37	33
4	40	36
5	41	39
6	45	44
7	46	47
8	49	51
9	50	53
10	53	56
11	55	57
12	57	62
13	60	63
14	62	64
15	65	64
16	66	64
17	69	64
18	70	65
19	71	65
20	71	65

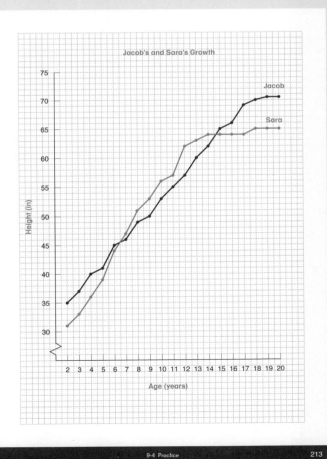

③ The following data shows the number of hours a week, to the nearest half hour, that some children spent playing organized sports.

$2\frac{1}{2}$	0	$1\frac{1}{2}$	5	0	2	$5\frac{1}{2}$	3	0
$9\frac{1}{2}$	$2\frac{1}{2}$	$8\frac{1}{2}$	$2\frac{1}{2}$	$3\frac{1}{2}$	8	5	$1\frac{1}{2}$	$5\frac{1}{2}$
$8\frac{1}{2}$	0	5	$3\frac{1}{2}$	8	3	6	4	3
3	$6\frac{1}{2}$	$2\frac{1}{2}$	9	5	$5\frac{1}{2}$	$2\frac{1}{2}$	$7\frac{1}{2}$	5

(a) Complete the line plot below.

Hours Playing Sports

Hours

(b) What were the two most common times spent playing sports?
  $2\frac{1}{2}$ and 5 hours

(c) What fraction of the students did not play any organized sports?
  $\frac{1}{9}$

(d) What fraction of the students played sports for more than 5 hours?
  $\frac{1}{3}$

Exercise 5 • pages 215–218

Exercise 5

Check

① Write the numbers in order from least to greatest.

(a) $\frac{7}{9}, \frac{13}{18}, \frac{3}{7}, \frac{15}{13}, \frac{8}{21}$

$\frac{8}{21}, \frac{3}{7}, \frac{13}{18}, \frac{7}{9}, \frac{15}{13}$

(b) $\frac{43}{7}, \frac{31}{5}, \frac{63}{8}, \frac{49}{9}$

$\frac{49}{9}, \frac{43}{7}, \frac{31}{5}, \frac{63}{8}$

② Find the values. Express each answer in simplest form.

(a) $3\frac{7}{12} - \frac{2}{3}$ $2\frac{11}{12}$

(b) $\frac{5}{8} + \frac{3}{4} + \frac{1}{2}$ $1\frac{7}{8}$

(c) $4\frac{1}{2} + 16\frac{5}{6}$ $21\frac{1}{3}$

(d) $7\frac{1}{9} - 5\frac{2}{3}$ $1\frac{4}{9}$

(e) $12 \times \frac{5}{9}$ $6\frac{2}{3}$

(f) $\frac{7}{8} \times 10$ $8\frac{3}{4}$

(g) $\frac{2}{7}$ of 98   28

(h) $39 \div 6$ $6\frac{1}{2}$

③ Find three common multiples of 2, 3, and 4 greater than 60 but less than 100.
The least common multiple of 2, 3, and 4 is 12. So the common multiples are multiples of 12.
72, 84, 96
Answers may vary.

④ Estimate and then find the quotient and remainder when the difference between 35,004 and 28,683 is divided by 6.
6,321 ÷ 6 is 1,053 R 3.

⑤ Estimate and then find the product when the sum of 4,789 and 589 is multiplied by 35.
5,378 × 35 = 188,230

⑥ Victoria has 30 pennies, 15 nickels, and 45 dimes.

(a) What fraction of her coins have a value of more than 1 cent?
$\frac{60}{90} = \frac{2}{3}$

(b) If she is given 10 quarters, what fraction of her coins will have a value less than 10 cents?
$\frac{45}{100} = \frac{9}{20}$

⑦ A jewelry store had 60 necklaces. They sold $\frac{3}{5}$ of them for $459 each and the rest for $295 each. How much did they receive from the sales?
$\frac{3}{5} \times 60 \times 459 = 16,524$ $\frac{2}{5} \times 60 \times 295 = 7,080$
16,524 + 7,080 = 23,604
$23,604

⑧ For an art project, Amy cut a 30 ft piece of yarn into different sized pieces. 10 pieces were $\frac{7}{12}$ ft long and 18 pieces were $\frac{3}{4}$ ft long. How long is the leftover piece of yarn in feet?
$10 \times \frac{7}{12} = 5\frac{5}{6}$
$18 \times \frac{3}{4} = 13\frac{1}{2}$
$30 - 5\frac{5}{6} - 13\frac{1}{2} = 10\frac{2}{3}$
$10\frac{2}{3}$ ft

⑨ Mr. Ikeda bought 5 identical lamps, 4 identical chairs, and 2 identical side tables. Each lamp cost half as much as each chair. Each side table cost the same as a lamp and chair combined. If he spent $248 on the chairs, how much did he spend in all?

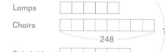

8 units ⟶ 248
1 unit ⟶ 248 ÷ 8 = 31
19 units ⟶ 19 × 31 = 589
$589

⑩ Avery had some blue and white beads. She used half of each type of bead. Then she had 3 times as many blue beads as white beads. How many times as many blue beads as white beads did she have at first?

Challenge

⑪ Aaron rolls a die. Every time he rolls a prime number, he gets 5 points. Every time he does not roll a prime number, he loses 3 points. At the end of the game, his score is 0. What is the least number of rolls he could have made? (Hint: The points he won and lost must be the same.)
Multiples of 5: 5, 10, 15
Multiples of 3: 3, 6, 9, 12, 15
Since 15 is the least common multiple, he would have rolled a prime number 3 times and would not have rolled a prime number 5 times.
3 + 5 = 8
8 rolls

⑫ There are 12 cards numbered 1–12. Alex, Dion, Mei, Emma, and Sofia each pick 2 cards and find the sum of the two cards they picked.

| 1 | 2 | 3 | 4 | 5 | 6 |
| 7 | 8 | 9 | 10 | 11 | 12 |

Alex	Dion	Mei	Emma	Sofia
16	4	19	11	7
6, 10	1, 3	8, 11	4, 7	2, 5

Which 2 cards did each friend pick?
Alex: 4 + 12 = 16, 5 + 11 = 16, 6 + 10 = 16, 7 + 9 = 16
Dion: 1 + 3 = 4
Mei: 7 + 12 = 19, 8 + 11 = 19, 9 + 10 = 19
Emma: 1 + 10 = 11, 2 + 9 = 11, 3 + 8 = 11, 4 + 7 = 11, 5 + 6 = 11
Sofia: 1 + 6 = 7, 2 + 5 = 7, 3 + 4 = 7
Dion had to have picked 1 and 3. So Sofia had to have picked 2 and 5.
1, 2, 3 and 5 have been picked, so Emma must have picked 4 and 7.
1, 2, 3, 4, 5, and 7 have been picked, so Alex must have picked 6 and 10.
The only ones left that Mei could have picked are 8 and 11.

Blackline Masters for 4A

All Blackline Masters used in the guide can be downloaded from dimensionsmath.com.
This lists BLMs used in the **Think** and **Learn** sections.
BLMs used in **Activities** are included in the Materials list within each chapter.

Equivalent Fractions	**Chapter 6:** Lesson 1
Equivalent Fractions Number Lines	**Chapter 6:** Lesson 1
Graph Paper	**Chapter 3:** Lesson 3, Lesson 4, **Chapter 9:** Lesson 3, Lesson 4
Hundred Chart	**Chapter 3:** Lesson 1, Lesson 2, Lesson 4
Line Graph	**Chapter 9:** Lesson 2
Number Cards	**Chapter 2:** Lesson 3
Number Line	**Chapter 1:** Lesson 5, Lesson 6
Place-value Cards	**Chapter 1:** Lesson 1, Lesson 2, Lesson 3, Lesson 4
Ship Weight	**Chapter 1:** Lesson 6
Truck Weight	**Chapter 1:** Lesson 5

Notes